ADVANCES IN MEDICINAL CHEMISTRY

Volume 5 • 2000

ADVANCES IN MEDICINAL CHEMISTRY

Editors: ALLEN B. REITZ
SCOTT L. DAX
Drug Discovery
The R. W. Johnson Pharmaceutical
Research Institute

VOLUME 5 • 2000

JAI PRESS INC.
Stamford, Connecticut

ISBN: 0-7623-0593-2
ISSN: 1067-5698

Transferred to digital printing 2006
Printed and bound by CPI Antony Rowe, Eastbourne

CONTENTS

LIST OF CONTRIBUTORS

Michael W. Decker

Neurological and Urological Diseases
Abbott Laboratories
Abbott Park, Illinois

Susan Hagen

Department of Chemistry
Parke-Davis Pharmaceuticals
Division of Warner-Lambert
Ann Arbor, Michigan

Gregory S. Hamilton

Department of Research
Guilford Pharmaceuticals, Inc.
Baltimore, Maryland

Mark W. Holladay

SIDDCO, Inc.
Tuscon, Arizona

Bradley D. Tait

Department of Chemistry
Parke-Davis Pharmaceuticals
Division of Warner-Lambert
Ann Arbor, Michigan

Christine Thomas

Department of Research
Guilford Pharmaceuticals, Inc.
Baltimore, Maryland

J. V. N. Vara Prasad

Department of Chemistry
Parke-Davis Pharmaceuticals
Division of Warner-Lambert
Ann Arbor, Michigan

David J. Wustrow

Department of Chemistry
Parke-Davis Pharmaceuticals
Division of Warner Lambert
Ann Arbor, Michigan

PREFACE

Volume 5 of *Advances in Medicinal Chemistry* contains four intriguing and detailed accounts of the close interface between synthetic chemistry, structure–activity relationships, biochemistry, and pharmacology. It is the confluence of these various disciplines that makes medicinal chemistry a thoroughly exciting field of endeavor.

Chapter 1 is a comprehensive survey of the immunophilin area specifically focusing on neuroregenerative applications in the central nervous system. Structure-based drug design is assisted by X-ray and NMR structural analysis.

Chapter 2 is an intriguing account of the development of a potent analgesic compound that works via modulation of neuronal nicotinic acetylcholine receptors. The clinical candidate ABT-594 can be traced to the unlikely origin of the natural product epibatidine from the skin of an Ecuadoran frog and a muscarinic receptor program geared towards treating Alzheimer's disease.

Chapter 3 describes dopamine D-2 autoreceptor partial agonists as potential therapy for the treatment of schizophrenia, a debilitating disease that takes a large toll on family and society. This chapter illustrates nicely the frustrations inherent in drug discovery research: new structural modifications that looked promising were eventually associated with some toxic or other untoward effect precluding continued development.

Chapter 4 tells of a successful program in which potent non-peptide inhibitors of HIV protease from the AIDS virus were developed. Coumarins and dihydropyranones were identified as viable chemical leads via screening. Rational drug design, aided by molecular modeling and X-ray crystal analysis, along with the development of sound structure–activity relationships, yielded a number of promising clinical candidates.

We thank all of the authors for the considerable time and effort that they have put forward in writing the chapters in this volume. We also acknowledge the R. W. Johnson Pharmaceutical Research Institute for continuing assistance and encouragement.

The R. W. Johnson Pharmaceutical Research Institute   Allen B. Reitz
Spring House, Pennsylvania   Scott L. Dax
                                                                              Editors

IMMUNOPHILINS:

THE NEXT GENERATION

Gregory S. Hamilton and Christine Thomas

Advances in Medicinal Chemistry
Volume 5, pages 1–84.
Copyright © 2000 by JAI Press Inc.
All rights of reproduction in any form reserved.
ISBN: 0-7623-0593-2

I. INTRODUCTION

In the course of pharmaceutical research, a large number of enzyme families have been extensively studied and utilized as targets for therapeutic drug design. In recent years, few classes of enzymes have had as interesting a history as the peptidyl prolyl isomerases (PPIases, or rotamases). Discovered as cellular catalysts for the isomerization of prolyl peptide bonds, the PPIases became a focus of intense interest when it was discovered that they were the cellular binding targets for important immunosuppressant drugs. The highly competitive effort to understand the PPIase-mediated mechanism of action of immunosuppression by these drugs comprised one of the major scientific detective stories in biochemistry in the late 1980s and early 1990s. In addition to elucidating the mechanism of action of the immunosuppressants, this work led to a large increase in knowledge about the PPIases.

II. PROLYL ISOMERASES AND IMMUNOSUPPRESSANT DRUGS[1]

The saga of the immunophilins begins in 1984, when the German biochemist Gunther Fischer isolated and characterized a new 18 kDa enzyme with a novel catalytic activity.[2] Because this enzyme catalyzed the interconversion of *cis*- and *trans*-amide bond rotamers in peptidyl-prolyl substrates (Figure 1), he named this enzyme peptidyl–prolyl *cis–trans* isomerase, or PPIase. Since isomerization about peptidylprolyl amide bonds is slower than for other residues and represents a potential rate-limiting step in protein folding and unfolding, the existence of an enzyme to facilitate this rotamerization was logical. Fischer demonstrated that PPIase did indeed accelerate the in vitro refolding of a denatured protein substrate.[3]

Also in 1984, a group led by R.E. Handschumacher was investigating the cellular actions of a new, extremely potent immunosuppressive drug named cyclosporin A (CsA). Cyclosporin A (Figure 2), a macrocyclic undecapeptide natural product produced by a fungus (*Tolypocladium inflatum*) from a Norwegian soil sample, showed extraordinary promise

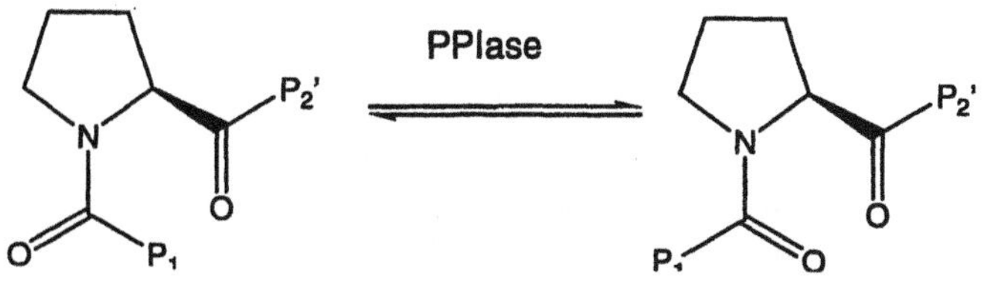

Figure 1. Peptidyl prolyl isomerases catalyze the interconversion of *cis*- and *trans*-amide bond rotamers adjacent to proline residues in peptidic substrates.

in organ transplant surgery to inhibit rejection of the transplanted tissue.[4] Handschumacher's group reported the isolation from calf thymus of a protein that was the principal binding protein for cyclosporin A, and named the protein cyclophilin (CyP).[5] Although the connection between these two events was not immediately made, by 1989 it was demonstrated that cyclophilin and PPIase were the same protein.[6,7] Thus was established the initial connection between peptidylprolyl isomerases and immunosuppressant drug action.

That same year brought another new peptidylprolyl isomerase onto the scene, also in the context of the mechanism of action of a novel immunosuppressive drug. The macrolide antibiotics FK506 and rapamycin had been identified as potent immunosuppressive agents isolated from *Streptomyces* fermentation broths.[8] FK506 was produced from a *Streptomyces* strain found in the volcanic soil at the base of Mount Fuji, *Streptomyces tsukubaensis*, while rapamycin was produced by *Streptomyces hygroscopius*, isolated from a soil sample from Rapa-Nui on Easter Island. FK506, in particular, was a considerably more potent immunosuppressive agent than CsA. Researchers at Merck Research Laboratories, and in Stuart Schreiber's group in the chemistry department at Harvard, reported at the same time the isolation of the major cytosolic binding protein for FK506.[9,10] Named FKBP (for FK506-binding protein), this 12 kDa protein had no sequence homology to cyclophilin. However, it too was shown to possess peptidylprolyl isomerase activity, and found also to be the binding target of rapamycin. While rapamycin and FK506 are close structural analogues of each other (Figure 2), cyclosporin A bears no resemblance to either, and cyclosporin binds to a PPIase (cyclophilin) with no sequence or structural homology to FKBP, the PPIase to which FK506 and rapamycin bind. In each case, the immunosuppressant drugs bind to the proline-binding site of their respective PPIases, and potently inhibit the PPIase or rotamase activity.

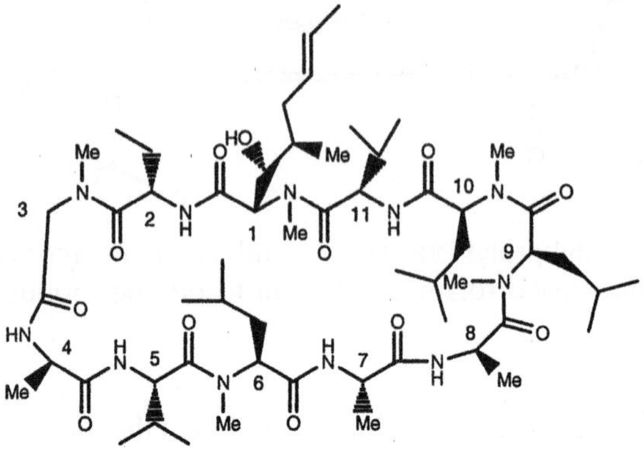

Cyclosporin A

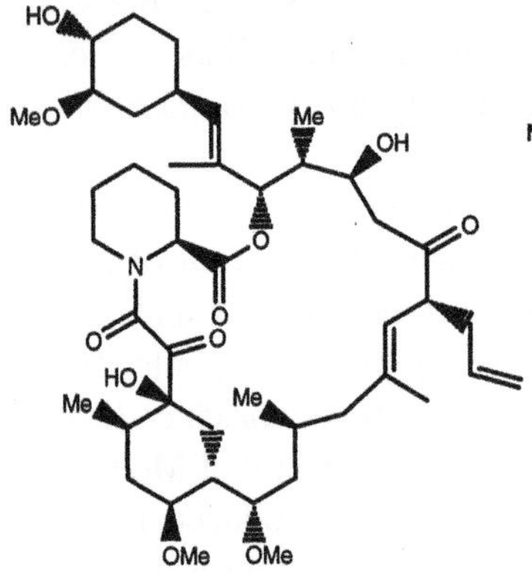

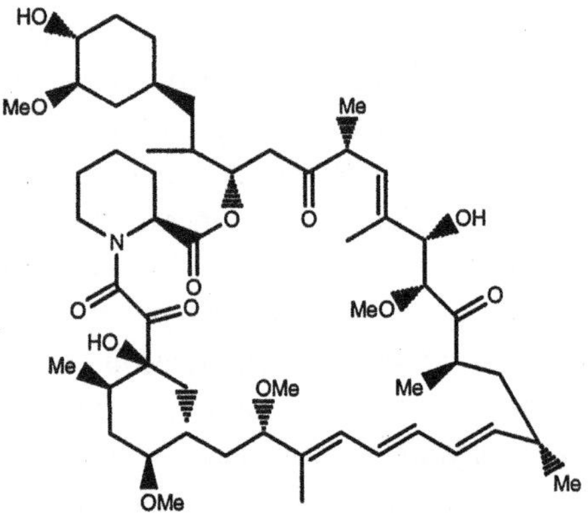

FK506

Rapamycin

Figure 2. Immunosuppressant drugs which work by an immunophilin mediated mechanism.

Because of the common property of the PPIases as targets for immuno-suppressant drugs, Schreiber renamed the PPIases the "immunophilins." This term has come to be the one most commonly applied to these proteins.

Thus, by 1989 it was known that the cytosolic target of each of the three immunosuppressive drugs was a peptidylprolyl isomerase. The mechanism of action of the drugs was at that time unknown, and an attractive hypothesis was that inhibition of PPIase activity was critical to the ability of the drugs to inhibit proliferation of T-lymphocytes (T-cells). However, there were problems with this hypothesis. FK506 and rapamycin appeared to have different mechanisms of action, although the two drugs were similar in structure and bound to the same immunophilin. Both FK506 and cyclosporin blocked early signaling pathways in T-cells associated with an increase in intracellular Ca^{+2} concentration, leading to the activation of T-cells and transcription of genes for interleukin-2 (IL-2) and its receptor.[11,12] Rapamycin, on the other hand, appeared to block later Ca^{+2}-independent T-cell events leading to cell cycle progression and proliferation.[13,14] Indeed, it was shown that FK506 and rapamycin are reciprocal antagonists of each other's immunosuppressive effects.[15] Schreiber thus proposed that, rather than FK506 inhibiting a function of FKBP, its PPIase activity, it promoted a *gain of function* for FKBP.[13] This was termed the "activated complex" hypothesis.

In a seminal piece of work, Schreiber's group synthesized 506BD (FK506 binding domain) that contained the α-ketoamide pipecolinyl portion common to both rapamycin and FK506 but which was truncated in the macrocyclic ring portion (Figure 3).[16,17] This compound was observed to be a potent inhibitor of FKBP PPIase activity but inactive as an immunosuppressant in vitro, suggesting that PPIase inhibition was insufficient for immunosuppression. Schreiber formulated a view in which the immunosuppressant drugs possessed two distinct domains: an immunophilin-binding domain, which bound to the rotamase active site of the cognate immunophilin, functioning as a high-affinity ligand and enzyme inhibitor, and an "effector" domain. The effector domain of the drugs, which extended beyond the surface of the immunophilin into the solvent accessible region, was postulated to be responsible for mediating the immunosuppressant effects of the immunophilin–drug complex.

Additional analogues of the immunosuppressant drugs which bound to their respective immunophilins but failed to suppress T-cell proliferation were subsequently reported by others. Several are depicted in Figure 3. Ascomycin is the C-21 ethyl analogue of FK506 and is immunosuppressant. Workers at Merck demonstrated that placing a hydroxyl group at C-18, to yield 18-hydroxyascomycin (or L-685,818), abolished immunosuppression,[18] suggesting that the immune system activity of FK506 is exquisitely sensitive to modifications in this region of the molecule. Nonimmunosuppressive effector domain-modified analogues of rapamycin and cyclosporin were also identified. Workers at Wyeth–Ayerst reported the preparation and characterization of WAY-124,466, a ra-

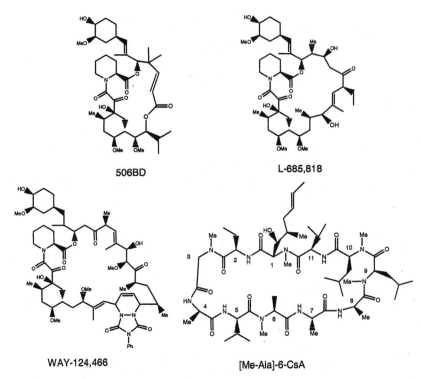

506BD

L-685,818

WAY-124,466

[Me-Ala]-6-CsA

Figure 3. Nonimmunosuppressive immunophilin ligands used to probe the mechanism of action of FK506, rapamycin, and cyclosporin A.

pamycin derivative modified in the triene portion of its effector domain which no longer blocks T-cell proliferation but retains FKBP affinity and rotamase inhibitory activity.[19] Me-Ala6-CsA is a cyclosporin analogue in which one of the amino acid residues in the cyclosporin effector domain has been modified, resulting in a compound which is a potent cyclophilin inhibitor but is not immunosuppressive.[20]

With the demise of the rotamase hypothesis of immunosuppression, an intensive effort commenced in several laboratories to discover the putative secondary protein targets of the drug–immunophilin complexes. In 1991, Schreiber's group at Harvard and Irving Weissman's group at Stanford identified the enzyme calcineurin as the target of both the FK506-FKBP and CsA-CyP complexes, using FKBP and CsA fusion-protein affinity columns.[21,22] Calcineurin, also known as phosphatase 2B, is a Ca^{+2}/calmodulin-dependent serine/threonine phosphatase.[23] The FK506-FKBP and CsA-CyP complexes were potent inhibitors of calcineurin's enzyme activity; however, neither FK506 or CsA, nor FKBP or CyP, was by itself an appreciable calcineurin inhibitor.

The discovery of calcineurin as a target for the drug–immunophilin complexes suggested a new hypothesis for the mechanism of immunosuppression: inhibition of calcineurin activity resulting in hyperphosphorylation of one of its substrates. Gerald Crabtree, an immunologist at Harvard, demonstrated that nuclear association of certain transcription factors in T-cells was blocked by FK506 and cyclosporin.[24] One of these transcription factors was nuclear factor of activated T-cells (NF-AT), which regulates transcription of the gene for IL-2. Since the cytosolic form of NF-AT, which is phosphorylated, cannot enter the nucleus, dephosphorylation by calcineurin activates the IL-2 promoter.[25] Thus in the presence of FK506 or CsA, calcineurin inhibition results in blockade of NF-AT translocation into the nucleus and subsequent IL-2 production.[26] Other transcription factors affected by CsA and FK506 by this route include the NF-kB factors and Oct-1/OAp.[24]

Shortly after the discovery of calcineurin as the mechanistic key for FK506 and CsA action, the mechanism of action of rapamycin began to unfold. It had already been noted that rapamycin blocked the IL-2 stimulated G1 to S phase transition in T-cells, inhibiting cell division. Treatment of T-cells with rapamycin was found to result in decreased enzymatic activity of several kinases, including p70 S6 kinase (a 70 kDa protein which phosphorylates the S6 protein of the small ribosomal subunit),[27–29] and cyclin-dependent kinases of 33 and 34 kDa.[30,31] However, in vitro experiments demonstrated that these kinases were not directly inhibited by the FKBP–rapamycin complex. In 1993, two yeast proteins were identified that appeared to be involved in the mechanistic pathway and mutations in these proteins conferred resistance to rapamycin-induced cytotoxicity.[32] These proteins were named TOR1 and TOR2 (targets of rapamycin).

In 1994 two groups independently and concurrently reported the identification of a protein which interacted with FKBP12 only in the presence of rapamycin. One of these groups was Schreiber's, who isolated the human protein and named it FRAP, for FKBP and rapamycin associated protein.[33] The other group, led by Solomon Snyder at Johns Hopkins, isolated a similar protein from rat brain extracts. They named this protein RAFT, for rapamycin and FKBP target.[34] Sequencing demonstrated that these were the mammalian homologues of the yeast TOR proteins.

Like the yeast homologue TOR2, RAFT/FRAP possesses a serine/threonine kinase domain with homology to phosphatidylinositol-3-OH kinase (PI3K). Schreiber's group used mutant forms of FRAP that did not interact with FKBP12-rapamycin to probe the relation between FRAP and S6 kinase inhibition.[35] They found that FRAP regulated S6 kinase activity in vivo in a manner that was rapamycin-sensitive, and depended upon FRAP kinase activity. In vitro, FRAP autophosphorylation was blocked by the FKBP12-rapamycin complex. Phosphorylation of S6 is correlated with increased translation of certain mRNAs following stimulation of signaling pathways by growth factors.[36,37] Another connection between rapamycin activity and regulation of translational pathways was made when it was discovered that rapamycin potently inhibits the phosphorylation of 4E-BP1. This protein, when dephosphorylated, binds to the transcription initiation factor eIF4E and inhibits the initiation of translation.[38] No direct interaction of FRAP with either p70 S6 kinase or 4E-BP1 has been observed, so it seems likely that additional, as yet unknown components of the pathway remain to be discovered.

Altogether, the story of the immunosuppressant drugs and the immunophilins has been a remarkable odyssey of discovery. Perhaps of equal importance is the role of immunophilin research in ushering in a new era in cell biology research. The immunosuppressant drugs have been extraordinarily fruitful as tools to probe previously unknown aspects of signal transduction pathways in T-cells. This work has shown in stunning fashion the power of synthetic chemistry as a tool for the study of cell biology. At all key junctures of the study of the actions of FK506, CsA, and rapamycin, experiments conceived and executed by chemists working together with biologists, utilizing rationally designed protein ligands, were crucial. The study of the immunophilins has been an incubator for a new field, which Schreiber terms "chemical biology," in which fundamental research in cell biology occurs at the interface of the disciplines of chemistry and biology. In the past, studies of protein function in cellular pathways were largely done using molecular biology techniques to generate mutant proteins and transfect them into cells for study. The increasing maturity of protein structure determination methods, structure-based drug design, and combinatorial chemistry suggests that design of selective ligands will increasingly be a major tool to study protein function in cells, and that the techniques and concepts of organic chemistry will be as important as those of molecular biology in elucidating the molecular mechanisms of cellular processes.

III. THE PEPTIDYLPROLYL ISOMERASE FAMILY OF PROTEINS

Since the discovery of the first two peptidylprolyl isomerases, FKBP12 and cyclophilin A, a large number of related PPIases have been discovered and studied, and it is now clear that the rotamases are an extremely large, ubiquitous, and highly conserved family of proteins widely found in both prokaryotic and eukaryotic organisms.[39,40] In mammals, rotamases are present and abundant in all tissues and in the majority of subcellular compartments.[40] Over a dozen different rotamases are present in humans and widely distributed in various tissues. Most of the rotamases identified to date are immunophilins related to either FKBP12 or cyclophilin A. A third class of PPIase now appears to be emerging unrelated to the immunophilins. The enzyme parvulin from *E. coli*, discovered in 1994, possesses rotamase activity comparable to cyclophilin but does not bind either FK506 or CsA.[41] A number of proteins homologous to parvulin occur in *E. coli* and other microorganisms, suggesting the existence of a third rotamase family.[42]

A. Enzymatic Activity of the Rotamases

Early on, the PPIases were divided into two classes, the FKBPs and cyclophilins, on the basis of their ability to bind either FK506 or cyclophilin. Members of these two classes are commonly called immunophilins, though this term is not meant to imply that all of them are involved in immunosuppressant drug action. As discussed above, other prolyl isomerases have been identified which do not bind known immunosuppressant drugs, suggesting significant structural divergence from the active sites of the FKBPs and cyclophilins. As will be described in the following section, structural analysis of immunophilins complexed with immunosuppressant drugs has been a powerful tool for delineating the prolyl isomerase domains in these two groups.

More than 20 FKBPs and at least 30 cyclophilins have been identified in recent years. New PPIases continue to be discovered with each passing year. We will focus on the better known and characterized immunophilins in mammals and their relevance to pharmaceutical research. We will briefly discuss the enzymatic activities and cellular locations of the major FKBPs and cyclophilins, and then move on to a detailed discussion of the structural biology of the immunophilins.

By convention, members of the FKBP family are named by appending to the prefix FKBP the apparent weight of the protein in kilodaltons. Thus well-known FKBPs include, in addition to FKBP12, FKBP12.6, FKBP13, FKBP25, FKBP38, and FKBP52. Unfortunately no rational system exists for cyclophilins. The first four discovered are named CyP-A, B, C, and D. On the other hand, a 40 kDa cyclophilin is called CyP-40, and a cyclophilin found in natural killer cells is denoted CyP-NK! Table 1 summarizes data for the major FKBPs and cyclophilins found in humans.

Assessment of enzymatic rotamase activity of the immunophilins is typically done using a spectrophotometric assay.[2] This assay is based on the fact that α-chymotrypsin is capable of liberating *para*-nitroanilide (pNA) from peptides containing the sequence Xaa-Pro-Phe-pNA only when the prolyl amide bond is in the *trans* conformation. When α-chymotrypsin is added to a solution of such a peptide, rapid release of pNA from the *trans* conformer population occurs which can be followed spectrophotometrically. Subsequent release of additional pNA is contingent upon the slow isomerization of the *cis* conformer population to the *trans*, and this process is accelerated by PPIase. The assay is complicated by the fact that at equilibrium in solution, greater than 90% of the substrate is already in the *trans* conformation, resulting in poor signal-to-noise ratios. Rich et al. have reported modified versions of the assay, using lithium salts and cosolvents to increase the proportion of *cis* conformation, which have facilitated the measurement of kinetic constants of numerous rotamases, and their substrate specificities.[60,61] Table 1 shows typical k_{cat}/K_m values reported in the literature for the various immunophilins, using this assay. Other means described for observing the rotamase activity of the immunophilins include the use of fluorescence or circular dichroism to follow protein folding catalyzed by PPIase activity,[6,62–64] and one- and two-dimensional NMR spectroscopy.[65,66]

Cyclophilin A itself is a catalytically efficient enzyme, with a measured k_{cat}/K_m of 10 μM^{-1}s^{-1} with the substrate tetrapeptide succinyl-Ala-Ala-Pro-Phe-pNA. It shows little selectivity towards residues in the P1 position, however. Harrison and Stein assayed a number of tetrapeptides as substrates for cyclophilin A and FKBP12.[67] Very little selectivity towards Xaa was found in the series succ-Ala-Xaa-Pro-Phe-pNA, except for a general preference for a hydrophobic P1 residue and a slight preference for Ala over others. In contrast, FKBP12 evinces a marked preference for branched alkyl residues at P1, particularly preferring leucine.

The mechanism whereby the rotamases catalyze the interconversion of amide bond rotamers has been a source of considerable interest. It was

Table 1. Properties of the Major Mammalian Immunophilins

Immunophilin	M_r	K_d nM	K_{cat}/K_m $\mu M^{-1}s^{-1}$	Cellular Localization	Reference	Comments
FKBP12	11,820	0.4 (FK506); 0.2 (rapamycin)	4.3 [h]	Cytosol	9, 10, 43, 44	
FKBP12.6		0.55 (FK506)	0.62 [h]		45, 46	
FKBP13	13,200	38 (FK506); 3.6 (rapamycin)	54 [y]	Endoplasmic reticulum	47	
FKBP25	25,325	160 (FK506); 0.9 (rapamycin)	0.8 [b]	Cytosol, nucleus	48, 49	
FKBP38		4.5 (FK506); 0.8 (rapamycin)			50	2 FKBP domains, 1 CyP domain
FKBP52	51,810	10 (FK506); 8 (rapamycin)	0.39 [h]	Nucleus, cytosol	51–53	3 FKBP domains, tetratricopep-tide domain
Cyclophilin A	17,737	2 (CsA)	10 [h]	Cytosol	9, 54	
Cyclophilin B	23,500	84 (CsA)	6.3 [h]	ER	55	
Cyclophilin C	22,795	4 (CsA)	—	Cytosol	56	
Cyclophilin D	19,981	3.6 (CsA)	0.9 [h]	Mitochondria	57	
Cyclophilin-40	[40,000]	300 (CsA)	1.9 [h]	Cytosol	58	
Cyclophilin-NK					59	

initially proposed that cyclophilin's activity involved formation of a tetrahedral adduct.[6] However, subsequent experimental results were inconsistent with this mechanism.[68,69] Schreiber's group synthesized radiolabeled FK506 and demonstrated that binding to FK506 was reversible and did not involve formation of a covalent adduct.[70] A mechanism of "catalysis by distortion," in which the enzyme stabilized the transition state in which the amide carbonyl was rotated 90° out of the amide group plane ("half-rotamased" or "twisted-amide" bond) was put forth. On the basis of their evaluation of a number of peptides as substrates for FKBP12, together with a structural analysis of the FKBP12-FK506 complex (see later), Schreiber's group proposed that FK506 and rapamycin function as transition state mimics of a Leu-Pro substrate fragment.[44] Because the keto carbonyl group of the α-ketoamide moiety is nearly orthogonal to the amide group in the ground state, this functionality is well suited for such mimicry.

Computational studies of the isomerization of tetrapeptide substrates by FKBP are consistent with this mechanism. Molecular dynamic and free-energy perturbation methods were used to study the binding and isomerization of the substrate N-acetyl-Ala-Ala-Pro-Phe-NMe.[71] The results suggested that the twisted transition state was stabilized by a combination of nonbonded interactions. The residues Trp-59 and Asp-37 were predicted to be particularly important, consistent with the mutagenesis results described in the next section. The results did support the notion that the ketoamide functionality acts as a twisted-amide surrogate; however the position of the Ala-Pro amide carbonyl oxygen in the transition state of these simulations resembled more closely that of the trans-amide carbonyl oxygen of FK506 or rapamycin than the keto carbonyl oxygen. The substrate was observed to bind in a turn configuration in both the beginning (cis conformer) and transition states. Similar results were found by Karplus et al., using molecular mechanics calculations augmented by ab initio results.[72] In this case the substrate N-acetyl-Ala-Leu-Pro-Phe-NMe was used. The substrate was predicted to bind as a proline turn; specifically, a type VIa β-turn. Structural analysis of an FK506-cyclic peptide hybrid bound to FKBP12 also suggests a β-turn recognition motif.[73]

Catalysis by distortion, preferential stabilization of the syn or twisted intermediate over the cis and trans forms, has also been proposed for cyclophilin.[68,69,74] The X-ray structure of N-acetyl-Ala-Ala-Pro-Ala-amidomethylcoumarin shows that this substrate binds also as a β-turn, but a type VIb rather than the type VIa suggested for FKBP.[75] It is suggestive that bradykinin and its Gly[6] analogue have been shown to be substrates for cyclophilin, the adoption of a β-turn by the C-terminal portion of bradykinin being well documented.[76] Similar to the proposed mechanism for FKBP, it has been suggested that cyclophilin rotamase activity is a combination of desolvation of the substrate induced by the hydrophobic pocket followed by stabilization of the *syn* intermediate.[74] Several complexes of small peptides with cyclophilin have been solved. In addition to the tetrapeptide noted above, these include Ala-Pro,[74] Ser-Pro, Gly-Pro, His-Pro,[77] succinyl-Ala-Pro-Ala-p-nitroanilide,[78] and succinyl-Ala-Ala-Pro-Phe-p-nitroanilide.[79,80] In all cases, the proline of the substrate binds in the *cis* conformation. However, in the recently solved structure of a complex between CyPA and a 25-mer from the HIV-1 gag capsid protein, the proline residue bound at the active site is in the *trans* conformation.[81]

B. The FKBP and Cyclophilin Families

FKBP12 and CyPA are the prototypical members of two large families of rotamases.[82] The residues which comprise the rotamase domain/FK506 binding site of human FKBP12, and the rotamase domain/CsA binding site of CyPA, define respectively an "FKBP domain" and "cyclophilin domain" which are remarkably conserved in the higher molecular weight family members and across species. Table 2 summarizes the conservation of these two domains in the immunophilins listed in Table 1. The three-dimensional structure of these immunophilin domains will be discussed in the next section.

Of the 13 residues that define the FKBP domain motif, 10 are highly conserved, and 7 (Tyr-26, Phe-36, Asp-37, Val-55, Ile-56, Tyr-82, and Phe-99) are completely conserved in all FKBPs, from all species, that possess significant rotamase activity. The position of greatest variability is His-87, which is replaced in other FKBPs by various hydrophilic or hydrophobic residues. Replacement of Phe-46 by other hydrophobic residues has little effect on rotamase activity.

FKBP12.6 has the same number of residues as FKBP12 and differs only by 18 mostly conservative changes.[45] The FKBP domain differs only by the replacement of Trp-59 by Phe. This change has little effect on either rotamase activity or FK506 affinity (Table 1), and the complexes of FKBP12.6 with FK506 or rapamycin are potent inhibitors of calcineurin or FRAP, respectively.[46] Messenger RNA for FKBP12.6 is abundant in the brain, and has been found in all tissues examined.

Table 2. Conservation of Immunophilin Domains in Human FKBPs and Cyclophilins

FKBP

Residue #	26	36	37	42	46	54	55	56	59	82	87	91	99
FKBP12	Tyr	Phe	Asp	Arg	Phe	Glu	Val	Ile	Trp	Tyr	His	Ile	Phe
FKBP12.6	•	•	•	•	•	•	•	•	Phe	•	Ala	•	•
FKBP13	•	•	•	Gln	•	Gln	•	•	•	•	Gln	•	•
FKBP25	•	•	•	Thr	Leu	Lys	•	•	•	•	Ser	•	•
FKBP52-d1	•	•	•	•	•	•	•	•	•	•	•	•	•
FKBP52-d2	Leu	•	•	•	Gly	Asp	Leu	Pro	Leu	Phe	Lys	•	Tyr
FKBP52-d3	•	Tyr	Glu	Asn	Gln	Arg	Leu	Ala	Leu	Leu	Asn	Leu	Leu

Cyclophilin

Residue #	55	60	61	63	72	101	102	103	111	113	121	122	126
CyP-A	Arg	Phe	Met	Gln	Gly	Ala	N	Ala	Gln	Phe	Trp	Leu	His
CyP-B	•	•	•	•	•	•	•	•	•	•	•	•	•
CyP-C	•	•	•	•	•	•	•	•	•	•	•	•	•
CyP-D	•	•	•	•	•	•	•	•	•	•	•	•	•
CyP-40	•	•	•	•	•	•	•	•	•	•	His	•	•
CyP-NK	•	•	•	•	•	•	•	Arg	•	•	His	•	•

FKBP13 was discovered shortly after FKBP12,[47] and is localized in the endoplasmic reticulum lumen.[83,84] Accordingly, it has a hydrophobic 21-residue sequence at the N-terminus which targets it to the ER, and a C-terminal –RTEL sequence which causes its retention in the pre-Golgi lumen. It is also widely distributed in all mammalian tissues examined. FKBP13 is less sensitive to FK506 and rapamycin than is FKBP12, and binds rapamycin with a 10-fold higher affinity than FK506, in contrast to FKBP12. The FKBP domain differs from FKBP12 by several nonconservative changes: Arg-42 and Glu-54 are replaced by Gln, and His-87 is replaced by Ala. As discussed in the next section, Arg-42 of FKBP12 interacts with FK506 but not rapamycin, so the R42Q substitution may contribute to the divergence in FK506 and rapamycin binding by FKBP13.

FKBP25 is even more selective (200-fold) for binding rapamycin over FK506.[48,49] FKBP25 is found in the nucleus as well as the cytosol of human T-lymphocytes,[85,86] and association of FKBP25 with DNA has been suggested. The N-terminal domain has been predicted to form the helix-loop-helix DNA-binding motif.[85] FKBP25 contains a sequence insertion not found in the other mammalian FKBPs. An additional seven amino acids (including a KKKK sequence) are inserted into the 40s loop (the loop between the two parts of the third strand of the β-sheet, which contains Arg-42 in FKBP12). These changes may account for the selectivity for rapamycin binding in FKBP25. Lysine, which repeats in the C-terminus, including the KKKK sequence in the 40s loop insertion, may serve as a nuclear localization sequence for FKBP25. Nine of the FKBP domain residues are conserved, the changes being R42T, F46L, E54K, and H87Q. Alone of the FKBPs, FKBP25 prefers Ala-Ala-Pro-Phe as a tetrapeptide substrate.

Some confusion in nomenclature exists for the immunophilin known variously as FKBP52, p59, FKBP59, and hsp56. Use of FK506- and rapamycin-affinity columns to isolate rotamase enzymes from T-cells resulted in the isolation of a high molecular weight immunophilin.[87] Although its migration on SDS/PAGE gel was indicative of a molecular weight of 56–59 kDa, calculation of its molecular weight subsequent to sequencing gave a value of 51.8 kDa.[51] This disparity between migration rates on gels and actual molecular weights is typical of the FKBPs. It was shown thereafter that the new FKBP immunophilin was identical to the previously described heat shock protein, hsp56, which is a component of unliganded androgen, estrogen, and glucocorticoid receptor complexes.[52,53,88] FKBP52 is widely expressed in vertebrate tissues which are responsive to steroid hormones.[51] FKBP52 is noteworthy as an example of an immunophilin with multiple immunophilin domains. It contains two FKBP domains with detectable prolyl isomerase activity.[89,90] A third FKBP domain has been postulated,[89] but as this putative domain has almost no sequence identity to other FKBP domains (Table 2), and no rotamase activity, its characterization as an FKBP domain seems dubious. Three tetratricopeptide repeat motifs are found in the C-terminal portion of the protein, followed by a calmodulin-binding consensus sequence. In the first FKBP domain (contained within residues 31–138), all but one of the rotamase domain residues are conserved, and this domain has PPIase activity comparable to FKBP12.[90] The second domain (within residues 148–253), has nine substitutions from FKBP12 (Y26L, F46G, E54D, V55L, I56P, W59L, Y82F, H87K, and F99Y). This domain does not appreciably bind FK506, and its rotamase activity is only 2% as large as the first domain. This domain also includes an ATP/GTP-binding site.[89] The putative third domain, within residues 267–373, has only one conserved residue and is inactive as a rotamase. FKBP52 is primarily found in the cytosol, but significant amounts exist in the nucleus, with the same pattern of distribution as hsp90 and steroid receptors.[91–93]

A 65 kDa immunophilin has been identified which contains four FKBP domains. FKBP65 is an ER protein with four domains homologous to FKBP13. Physical biochemical studies on chicken FKBP65 suggests that the four domains are arranged in a linear structure, and all are expected to be accessible to substrates.[94] However, only one domain is susceptible to inhibition by FK506. Oddly, this same domain appears to be inhibited by cyclosporin A. FKBP65 was observed to catalyze the refolding of type III collagen in vitro.

Two new immunophilins have recently been discovered which contain both FKBP and cyclophilin domains. A 37 kDa protein isolated from the Jurkat T-cell line was found to bind FK506, rapamycin, and cyclosporin A, all with high affinity.[50] FKBP37 contains two FKBP domains and one cyclophilin domain, and also possesses glyceraldehyde 3-phosphate dehydrogenase activity. The 52 kDa protein, isolated from human lymphoid cells, also binds all three immunosuppressant drugs but shows no detectable rotamase activity towards a variety of peptidic substrates.[95]

The cyclophilin rotamase domain is very highly conserved in the cyclophilin family. As shown in Table 2, cyclophilins A–D all have identical CyP domains. Human cyclophilin B, which is localized to the endoplasmic reticulum, contains an ER-directed sequence. Comparison of cyclophilin B, including human, yeast, *Drosophilia nina A*, and rat cyclophilin-like protein, with human cyclophilin A shows that all of them contain a conserved central rotamase domain which is flanked by variable N- and C-terminal domains.[55] The ER signal sequence is found in the N-terminal domain. All the cyclophilin B members have potent, CsA-sensitive rotamase activity. The CyPB-CsA complex ($K_i < 20$ nM) is a more potent inhibitor of calcineurin than is the CyPA-CsA complex ($K_i = 270$ nM).[96] In addition to its presence in the ER, cyclophilin B has also been detected in human milk[97] and in the blood.[98]

Cyclophilin C was first characterized in the rat, where it was observed to be expressed in a smaller subset of tissues than was CyPA or CyPB.[22] In humans CyPC is abundant in the kidney, pancreas, skeletal muscle, heart, lung, and liver, but is almost absent from T-cells and the brain.[56] CyPC binds CsA with an affinity comparable to that of CyPA. It also binds a 77 kDa glycoprotein, called cyclophilin-C-associated-protein (CypCAP) in the absence of CsA, which is a competitive inhibitor of this interaction.[22,99] The physiological relevance of the CyPC-CypCAP interaction is not known.

Cyclophilin D is a mitochondrial matrix protein whose role in mitochondrial permeability transition is well characterized.[57,100,101] Mitochondria in the heart, liver, and brain form large pores in response to oxidative stress and elevated Ca^{+2} levels. These pores are blocked by CsA. Studies with photoaffinity labeled CsA derivatives identified CyPD as a likely mediator of the pore blockade by CsA.[102]

Cyclophilins A–D all have conserved rotamase domains, and variation in enzymatic activity and CsA binding within this series is likely to be a function of changes in the flanking loop regions. In CyP-40, a 40 kDa cyclophilin initially isolated from calf brain,[58] divergence from the CyPA domains occurs. CyP-40 has considerably lower affinity for CsA than cyclophilins A–D, and this may be due to the W121H change in the binding domain.[103] On the basis of sequence alignments, CyP-40 appears to have two domains: the cyclophilin domain is located in the N-terminal portion of the protein, while the C-terminal region contains a tetratricopeptide repeat domain with homology to the C-terminal portion of FKBP-52.[104] Both FKBP-52 and CyP-40 have been shown to be components of steroid receptor complexes.[105]

A gene product specific for natural killer cells in mice and humans has been characterized as a PPIase of the cyclophilin family.[59] CyP-NK appears to be important in NK-cell cytotoxicity. This multiple domain protein contains a large hydrophobic domain in the N-terminus followed by the cyclophilin domain, with several additional hydrophobic domains following the latter. The role played by the CyP domain is not currently known.

C. Cellular Functions of the PPIases

As discussed in an earlier section, the immunophilins were first characterized as prolyl isomerases that could accelerate the refolding of proteins. This has been demonstrated for a number of substrates. In addition to ribonuclease A,[3] CyPA has been shown to catalyze the refolding of immunoglobulin chains,[62,106] carbonic anhydrase,[107] and RNAse T1.[62,108] CyPA also catalyzes the cis–trans isomerization adjacent to proline in bradykinin[76] and calcitonin.[109] FKBP has also been shown to catalyze refolding of carbonic anhydrase[110] and RNAse T1[111] in vitro, but its poor efficiency in this regard casts doubt on the in vivo relevance of these activities.

Immunophilins of both the FKBP and cyclophilin families have been observed to play increasingly important roles as components of macromolecular complexes and in protein trafficking. Both FKBPs and CyPs have been found to be involved in the translocation and maturation of proteins in the secretory pathway. Cyclophilins are required for the transport of properly folded rhodopsin from the ER to the cell surface in *Drosophila*.[112,113] They also appear to be involved in the folding and secretion of transferrin from hepatocytes,[114] and in the formation of triple-helical collagen in fibroblasts,[115] both of which processes take place in the endoplasmic reticulum. CsA inhibited the latter two processes but FK506 and rapamycin had no effect. Other ER-localized processes affected by CsA are the expression of functional nicotinic cholinergic and serotonergic-3 receptors,[116] suggesting that an ER-specific cyclophilin may be involved in maturation of these proteins. FKBP65 has been shown to be involved in the trafficking of tropoelastin,[117] and the trigger factor of *E. coli*, which is involved in protein translocation, contains an FKBP domain.[118]

Both CyP-40[105,119,120] and FKBP-52[88,121] are involved as chaperones in hsp90-dependent signal transduction pathways. These immunophilins bind to a common site on hsp90, a heat shock protein associated with steroid receptors.[122] Interaction of FKBP52 with hsp90 does not occur through its FKBP domain, but rather at the tetratricopeptide repeat domains. Since CyP-40 also contains tetratricopeptide domains, it may interact similarly. FKBP52 seems to not be required for heterocomplex formation, but plays a role in targeted movement of the receptor. FKBP65 forms complexes with hsp90 and the serine/threonine kinase c-Raf-1, implicating this immunophilin also in hsp90-mediated signal transduction pathways.[123]

Another heat shock protein–immunophilin chaperone complex is involved in the trafficking of cholesterol.[124] A macromolecular complex consisting of FKBP52, caveolin (a 22 kDa protein that plays a role in regulating cholesterol concentration), CyPA, CyP-40, and cholesterol, transports newly synthesized cholesterol from the ER to caveolae. In cells expressing the complex, treatment with either CsA or rapamycin disrupted the complex and interfered with the rapid transport of cholesterol.

Involvement of immunophilins with calcium-related cell signaling has emerged as a recurring theme. The ryanodine receptor (RyR) and the inositol 1,4,5-triphosphate receptor (IP_3R) are both calcium ion channels whose properties are regulated by FKBP12. The IP_3R receptor is found primarily in the endoplasmic reticulum, but also in cell membranes, and mediates calcium release from the ER or through the lipid bilayer, respectively.[125,126] Activation of cell surface receptors coupled to phospholipase C leads to generation of IP_3 and subsequent activation of IP_3Rs.[127,128] The receptor is a tetramer of four identical 320 kDa monomers.[129] Each of the four subunits of the IP_3 receptor is tightly associated with a molecule of FKBP12.[130] Dissociation of FKBP12 from IP_3R, by treatment with FK506 or rapamycin, results in increased calcium conductance through the channel, suggesting that the physiological role of FKBP12 in the complex might be to modulate the gating properties of the channel. Treatment of brain membranes with FK506 results in a decreased ability of the endoplasmic reticulum to accumulate calcium, and an order of magnitude increase in the calcium-releasing potency of IP_3.[130] It has been found that calcineurin associates with the IP_3R-FKBP12 complex and modulates Ca^{+2} flux by regulating phosphorylation of IP_3R.[131] FKBP12 may function as a docking protein to promote assembly of the multiprotein complex, since treatment of the heterocomplex with FK506 dissociates FKBP12 and calcineurin from the receptor. The FKBP12 binding site on the IP_3 receptor has recently been localized to Leu(1400)-Pro(1401).[132] In this same report, it was shown that mutant FKBP12s which lacked rotamase activity, but retained FK506 binding affinity, interacted robustly with IP_3R.

The situation with the ryanodine receptor is highly analogous. RyR is found in the sarcoplasmic reticulum and is involved in excitation–contraction coupling in muscle tissue.[133] High levels are also found in the heart and brain.[134] The ryanodine receptor is a tetramer of identical 565 kDa subunits.[135-137] Each of the four subunits of RyR tightly binds a molecule of FKBP; in skeletal muscle, FKBP12 is bound,[138] while FKBP12.6 complexes with the cardiac RyR.[46,139] Like IP_3R, the association with FKBP improves the conductance properties of the channel.[140] FKBP can be displaced from the receptor by FK506, with the result that calcium conductance is increased. This results in a diminished ability of the sarcoplasmic reticulum to accumulate calcium because Ca^{+2} which is pumped in leaks out through the receptor. FKBP12 stabilizes both the open and closed states of the ion channel so that the channel is more difficult to open in the presence of FKBP12, but once open calcium conductance is optimal.[141] Although the interaction with RyR appears to occur through the rotamase domain of FKBP12, rotamase activity per se is not necessary. Displacement of wild-type FKBP12 with mutant FKBP12s devoid of rotamase activity did not affect the calcium conductance properties of RyR.[142] The locations of FKBP12 and calmodulin (CaM) in the complex with the skeletal muscle RyR have recently been elucidated by electron cryomicroscopy.[143] Both FKBP and CaM bind to domains directly attached to a cytoplasmic extension of the transmembrane portion of the receptor. Structural changes induced in this domain might be coupled to alterations in the gating properties of the calcium channel.

The IP_3R and RyR are examples of FKBP12 functioning as an adapter or coupling molecule, assembling macromolecular complexes involved in cell signaling pathways. In its role of regulating the conductance properties of the receptors and their phosphorylation states, FKBP12 may modulate generation of calcium waves or oscillations.

Regulation of calcium currents by a cyclophilin has also been observed. Using the yeast two-hybrid system, a protein ligand for cyclophilin B was discovered which regulates calcium currents in T-cell signaling.[144,145] Called calcium signal modulating cyclophilin ligand, or CAML, this protein acts downstream of the T-cell receptor but upstream of calcineurin. CAML may be an evolutionarily conserved signaling molecule, since homologues have been found in other species.[146] FKBP12 also plays a role in calcium signaling in immune cells; it is released by mast cells and stimulates calcium signaling in polymorphonuclear granulocytes (PMNs).[147] These findings, together with the observation of chemokine-like activity by cyclophilin secreted from stimulated cells,[148,149] suggest a multifaceted role for the immunophilins in immune system pathways. Cyclophilin A has been detected in synovial fluids from sufferers of rheumatoid arthritis, and appears to function as a chemotactic agent for polymorphonuclear granulocytes.[150] Both FKBP12 and cyclophilin A are released by fibroblasts during apoptosis,[151] and coregulation by cyclophilins of histamine expression and biosynthesis of the leukotriene C4 has been reported.[152]

Another immunophilin–protein interaction discovered by screening a cDNA library in the yeast two-hybrid system is the association of FKBP12 with the transforming growth factor-β type I receptor (TGF-βRI). Transforming growth factor-β (TGF-β) binds to membrane-bound heterodimeric receptors.[153] The monomeric units are denoted type I and type II (TGF-βRI and TGF-βRII, respectively), and possess serine/threonine kinase activity. Binding of TGF-β requires the RI subunit, which also appears to initiate the signal. Transduction of the signal does not occur in the absence of TGF-βRII, however, suggesting that binding of TGF-β to the dimer causes TGF-βRII to activate TGF-βRI.[154] Wang et al. found that FKBP12 associates specifically with the unliganded TGF-β type I receptor, and that this interaction was competitively inhibited by FK506.[155] They further demonstrated that binding of TGF-β induced phosphorylation of TGF-βRI by TGF-βRII, with concomitant release of FKBP12 from TGF-βRI.[156] Phosphorylation of the type I receptor occurs in a glycine- and serine-rich region called the GS box, and FKBP12 binds to this same region. Displacement of FKBP12 from TGF-βRI by FK506 or nonimmunosuppressive analogues enhanced TGF-β-elicited functional responses. A mutant form of FKBP12 which binds FK506, but lacks calcineurin inhibitory ability (the G89P, I90K double mutant),[157] was incapable of inhibiting TGF-β signaling. Wang et al. proposed a model whereby association of FKBP12 with the type I receptor keeps the receptor in an inactive state.[156] Ligand-induced dimerization causes the type II receptor to phosphorylate type I, activating it by releasing FKBP12. The result with the calcineurin-inactive mutant suggests that FKBP12 may function to anchor calcineurin to TGF-βRI, keeping it in an inactive dephosphorylated form. In cells expressing mutant forms of TGF-βRI that do not bind FKBP12, signal transduction induced by TGF-β is unimpeded.[158–160] These type I receptor mutants which are deficient in FKBP12 binding have an enhanced tendency for spontaneous activation in the absence of ligand, consistent with the hypothesis that TGF-βRI activity is negatively regulated by FKBP12. Further support is provided by Schreiber et al., who constructed fusion proteins containing the inactive double mutant of FKBP12 and the intracellular domain of TGF-βRI.[161] A myristoyl tail was included on the mutant FKBP12 to target the fusion protein to the plasma membrane. In cells overexpressing the myristoylated cytoplasmic tails of TGF-βRI and TGF-βRII, signaling was constitutive, but this autosignaling was inhibited when the fusion protein containing FKBP12 was expressed. In the latter system, signaling could be restored by treatment with rapamycin or an FK506 derivative. The binding of FKBP12 to TGF-βRI was localized to a Leu-Pro sequence, as was the case for IP$_3$R.

As described previously, FKBP13 contains a signal sequence for retention in the ER. FKBP13 is upregulated in response to heat shock and accumulation of misfolded proteins in the ER, suggesting a chaperone role for folding of proteins in the ER.[162,163] A non-ER role for FKBP13 was recently discovered, however. Using the yeast two-hybrid system, Walensky et al. reported that FKBP13 associates with a homologue of Band 4, the erythrocyte membrane cytoskeletal protein.[164] Named protein 4.1G, this homologue of 4.1 is much more widely expressed than 4.1R, and interacts with FKBP13 at its carboxy-terminal domain. A His-Pro sequence at the carboxy terminus was shown to be necessary for FKBP12 binding. Experiments with antibodies against FKBP13 suggested that FKBP13 was present in red blood cells (which lack an ER), supporting a non-ER role for this immunophilin.

Several published studies provide some insight into the developmental role of immunophilins by analyzing the temporal profile of immunophilin expression. In chick embryos, FKBP12 was expressed by day 4 and was primarily associated with cardiomyocytes and osteo-chondrocytes.[165] FKBP12 levels in the heart increased as embryonic development progressed. FKBP12 expression correlated with mitosis in proliferating cells but not with DNA synthesis. The results suggest that FKBP12 may be important in cardiac development, and may function by affecting cellular Ca^{+2} levels. A recent report on FKBP12 knockout mice is consistent with this.[166] Shou et al. reported that mutant mice deficient in FKBP12 had normal skeletal muscle but exhibited severe cardiac defects, including severe dilated cardiomyopathy and ventricular septal defects.

In situ hybridization experiments revealed that FKBP52 mRNA is overexpressed in rabbit and rat testes, relative to other organs.[167] The expression of FKBP52 was found to be highly specific to certain stages of development and cell types during male germ cell differentiation.

Sananes et al. suggest that the immunophilin plays a specific role in the cell division process.

Human cyclophilins A, B, and C have sequence homology to NUC18, a nuclease involved in apoptosis, and have been shown to possess nucleolytic activity, suggesting they may play a role in apoptotic degradation of DNA.[168] Cyclophilins can degrade both single- and double-stranded DNA, and their nucleolytic activity is stimulated by Ca^{+2} and/or Mg^{+2}. The nucleolytic activity of cyclophilins is not inhibited by CsA, indicating that it is independent of PPIase activity.

Formation of infectious virion particles of human immunodeficiency virus-1 (HIV-1) requires cyclophilin A acting as a chaperone. CyPA binds to the HIV-1 gag protein,[169] a polyprotein which is cleaved by proteases to yield the virion structural proteins, including matrix, capsid, and nucleocapsid proteins. Franke et al. demonstrated that the interaction with cyclophilin A was mediated by a conserved proline rich region of the capsid protein, and that one of these prolines (Pro222) was required for both CyPA interaction and virion infectivity.[170] Treatment with cyclosporin A,[170,171] or nonimmunosuppressive analogues of cyclosporin,[171,172] blocked the CypA–virion association and reduced virion infectivity, suggesting that the immunophilin–virion association is physiologically relevant. Subsequent reports have provided additional information on the requirement of cyclophilin for HIV-1 infectivity and replication[173,174] and its interaction with the gag protein. The X-ray crystal structure of CyPA bound to the amino-terminal domain of the HIV-1 capsid protein,[175] and to a 25-amino acid peptide fragment of capsid,[81] have been reported. In both cases, the proline bound in the cyclophilin pocket (Pro-90 of the capsid protein) adopts the *trans* rotamer conformation, in contrast to the di-, tri- and tetrapeptide complexes described previously. The structure is another suggestion that the immunophilins may act both as sequence-specific binding proteins as well as relatively nonspecific prolyl isomerases. A recent report that both FKBP12 and cyclophilin can interact with the V3 loop of the HIV-1 coat protein (gp120) also suggests a role for immunophilins in HIV pathogenesis.[176]

As the brief review above demonstrates, immunophilins play roles in a wide variety of cellular processes. In a subsequent section, we will discuss in some detail the presence and therapeutic utility of immunophilins in the mammalian nervous system. First, however, we will describe the structural biology of the PPIases.

IV. STRUCTURAL BIOLOGY OF THE IMMUNOPHILINS AND THEIR LIGAND COMPLEXES

In order to study the biological function and interaction of macromole-cules, a detailed structural characterization of the protein and its sub-strates must be available. Two experimental techniques used to determine three-dimensional structure are X-ray crystallography and nuclear mag-netic resonance (NMR) spectroscopy. Computational modeling is an-other tool, most commonly used in conjunction with one of the above methods. Because the immunophilins were implicated in the mechanism of action of immunosuppressant drugs early on, and the various immu-nosuppressant drugs provided excellent selective and high-affinity li-gands for FKBP12 and cyclophilin A, three-dimensional structures of the free proteins and their drug complexes were avidly pursued.[177] These structures have provided a wealth of understanding of the structural biology of the two classes of immunophilins, and have provided a structural basis for understanding the increasingly expanding role of these fascinating proteins in cellular processes. We will review in the next two sections the current knowledge of structure–function relationships in the immunophilins.

The structures of unliganded FKBP12 and human cyclophilin A have been determined by X-ray crystallography and NMR spectroscopy, as have their complexes with FK506 and cyclosporin A, respectively (Fig-ure 4). These protein–ligand complexes have been invaluable for identi-fication and analysis of the proline-binding rotamase sites of the two archetypal immunophilins. These two respective prolyl isomerase do-mains—a "FKBP domain" and a "cyclophilin domain"—are observed to be well-conserved throughout the expanding family of FKBPs and cyclophilins. It was noted earlier that computational studies of the enzymatic mechanism of the PPIases suggested that FKBP12 and cyclo-philin may recognize different types of β-turn. Comparison of the two domains bound to ligands suggests that they recognize a common ele-ment, proline, in different configurational contexts. These two domains, FKBP12 and CyP-A, represent two different motifs, variations on which occur in the two immunophilin families.

In the subsequent discussion of the structural biology of the immuno-philins, all three dimensional structures are from the Brookhaven Protein Data Base, unless otherwise noted. Accession numbers for all protein structures are given at the end of the chapter. All graphical analysis was done using the modeling program Sybyl and associated modules.[178]

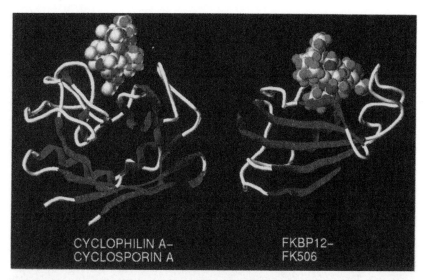

Figure 4. X-ray structures of cyclosporin A bound to cyclophilin A (*left*) and FK506 bound to FKBP12 (*right*). The proteins are depicted in ribbon format, color-coded by secondary structure, and the ligands are shown in space-filling representation.

A. FKBPs

The structures of FKBP12 (in complex with FK506), FKBP25 (in complex with rapamycin), and FKBP52 (the N-terminal domain) are on file at the Brookhaven Protein Data Bank. The protein structures and rotamase domains of these three FKBPs, *sans* ligands, are depicted in Figure 5 for comparison and reference in the following discussion of the FKBPs.

FKBP12

The structures of unliganded human[179] and bovine[180] FKBP12 were determined some years ago by X-ray crystallography and NMR spectroscopy. FKBP12 comprises a five-stranded antiparallel β-sheet wrapped around a short α-helical segment, with a right-hand twist (Figures 4 and 5).[179] The β-sheet is composed of residues 1–7, 20–29, 46–49, 71–77, and 96–107, and the α-helix, consisting of residues 59–65, is connected to strands five and two by intervening flexible loops. The topology of the β-sheet (+3, +1, −3, +1) is unusual in that it requires crossing between two of the loops, residues 8–19 and 50–70, the latter

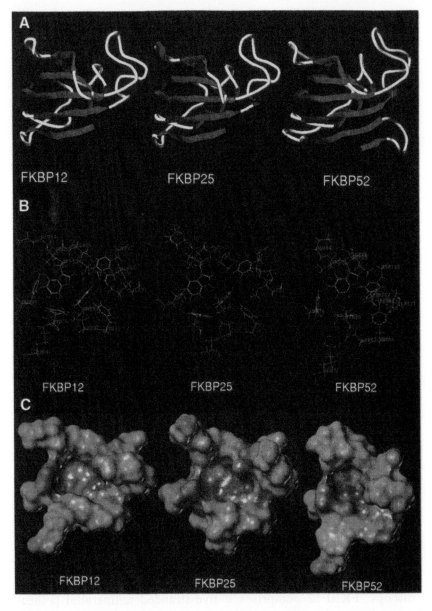

Figure 5. Comparison of FKBP12, FKBP25, and N-terminal domain of FKBP52. (**A**) Ribbon structures of the three FKBPs. (**B**) Rotamase domains of the FKBPs. (**C**) Lipophilic potential mapped onto solvent-accessible (Connolly) surfaces of the three rotamase domains. (**D**) Electrostatic potential mapped onto the solvent-accessible (Connolly) surfaces of the three rotamase domains.

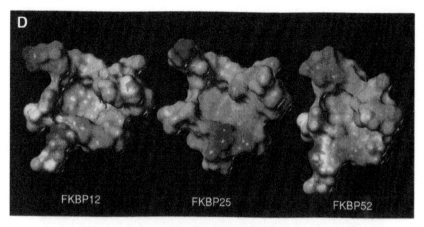

Figure 5. Continued.

of which contains the α-helix. Prior to the determination of the FKBP12 structure, loop crossings had been unobserved in antiparallel β-sheets and were thought to be forbidden.[181] The structure of FKBP12 demonstrates that it is possible to satisfy the hydrogen bond demands of the backbone amides in the two segments of such a crossing by a series if inter- and intrastrand hydrogen bonds.[182] The rotamase domain is a cavity between the α-helix and the β-sheet, and is lined with hydrophobic residues and flanked by loops 39–46, 50–56, and 82–95 (Figure 6). The first and third loops (denoted the 40s and 80s loops, respectively) are particularly flexible in solution in the unliganded protein.

The structure of the complex between human FKBP12 and FK506 has been solved by X-ray crystallography and NMR spectroscopy.[182–184] The pipecolinyl and pyranose rings of FK506, and ketoamide linkage, together with a portion of the cyclohexylpropenyl ester side chain, bind deeply into the pocket. The remaining portion of the molecule protrudes beyond the surface of the protein. A set of mostly hydrophobic residues interact with FK506 and define the binding domain: Tyr-26, Phe-36, Asp-37, Arg-42, Phe-46, Glu-54, Val-55, Ile-56, Trp-59, Tyr-82, His-87, Ile-91, and Phe-99 (Figure 6). Two hydrophobic pockets formed by these residues are occupied by FK506. The proline-binding pocket is formed by Trp-59 of the α-helix and the side chains of Tyr-26, Phe-46, Phe-99, Val-55, and Ile-56, and is occupied by the pipecolinyl ring of FK506. The adjacent site, occupied by the pyranose ring, is a cavity formed by the sidechains of Ile-91, His-87, Phe-36, Tyr-82, and Asp-37. Hydrogen bonds are formed between the ester carbonyl of FK506 and the back-

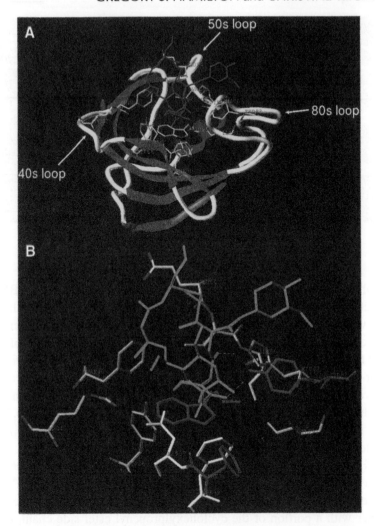

Figure 6. FK506-FKBP12 interactions, from the X-ray structure. (**A**) FK506 (red) in binding pocket surrounded by 40s, 50s and 80s loops of FKBP12. FKBP12 is color-coded by secondary structure. (**B**) FK506 (cyan) sur-rounded by residues comprising the rotamase domain of FKBP12. Residues important for both ligand binding and rotamase activity (Asp-37, Phe-99) are colored yellow; residues important for enzymatic activity but not ligand binding (Trp-59, Phe-36, Tyr-82) are colored red. Hydrogen bonds between FK506 and Ile-56 and Tyr-82 of FKBP12 are shown as yellow dashed lines. (**C**) FK506 (blue) surrounded by hydrophobic pockets of the FKBP12 rotamase domain. Solvent accessible surfaces formed by the residues Tyr-26, Phe-46, Val-55, Ile-56, Trp-59, and Phe-99 (colored green), and Phe-36, Asp-37, Tyr-82, His-87, and Ile-91 (colored red) are shown.

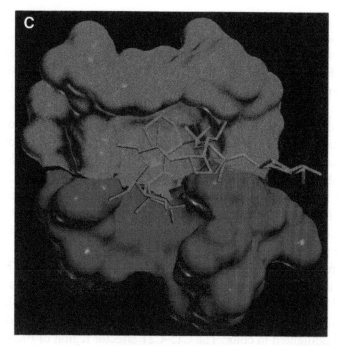

Figure 6. Continued.

bone—NH of Ile-56, and between the side chain—OH of Tyr-82 and the amide carbonyl of the ketoamide moiety. The cyclohexylpropyl ester chain of FK506 lies in a shallow hydrophobic groove on the protein surface. Figure 6 depicts the solvent-accessible protein surface of these two subsites of the FKBP domain.

In addition to FK506, the structures of a number of other macrocyclic and nonmacrocyclic ligands bound to FKBP12 have been solved by X-ray or NMR. These include the C-21 ethyl analogue of FK506, ascomycin,[185] rapamycin,[182,186] the nonimmunosuppressive analogue L-685,818 (18-hydroxyascomycin),[187] and several small molecule inhibitors.[188–190] In all of these structures, the FKBP-binding portions of the molecules are closely superimposable and partake of similar interactions. The solvent accessible portion of the macrocyclic molecules is crucial for mediating the interaction with the secondary protein target molecules (calcineurin or FRAP). As seen by the example of L-685,818, even very minor modifications of this effector element can disrupt this interaction and vitiate the immunosuppressive activity of the drug–immunophilin complexes.

Mutagenesis studies on both FKBP12 and CyPA have demonstrated that PPIase activity, ligand-binding, and immunosuppressant activity may be structurally dissected. In studies on the binding domain residues of FKBP12, the mutations F36Y[191] and Y82L[192] caused significant loss of enzymatic activity with much less effect on macrolide binding, whereas changing Trp-59 to Ala[193] abolished PPIase activity but had little effect on FK506 binding. Other changes (D37V,[194] F99W[142]) abolished both rotamase and ligand binding activities. In Figure 6, the residues of the domain are color-coded according to the mutagenesis results. Single- and double-mutant studies indicated that Asp-37, Arg-42, His-87, Gly-89, and Ile-90 were important in calcineurin binding by the composite surface formed by FKBP12 and FK506.[192,193] Arg-42 and Gly-89/Ile-90 are part of the 40s and 80s flexible loops, respectively, surrounding the binding pocket.

The recently solved X-ray crystal structures of the FKBP12-FK506–calcineurin ternary complex provide a detailed picture of these interactions.[195,196] Figure 7 depicts the FKBP12–FK506 complex in space-filling representation, with the key calcineurin-binding residues of FKBP highlighted in color. The C15-C21 effector region of FK506 is a complementary surface to a cleft in calcineurin; the allyl group of the immunosuppressant penetrates deeply into a pocket on the surface of the phosphatase. The regions of FKBP12 noted in Figure 7 (Asp-37–Asp-41, Arg-42–Phe-46, and His-87–Ile-90) also contact calcineurin. The FKBP12–FK506 complex does not bind at the enzymatic active site of calcineurin, but lies about 10 Å distant. Rather than directly inhibiting the phosphatase activity, binding of the complex appears to hinder approach of substrates to the active site.

The ternary complex of FKBP12/rapamycin/FRAP from X-ray data has been reported by Schreiber et al.[197] There do not appear to be any extensive protein–protein interactions in the ternary complex like that found in the FK506–FKBP12–calcineurin complex. The overall structure of FKBP12 does not differ significantly in comparison with the binary complex. Deviation in the conformation in the 80s loop portion is evident around Ile90, suggesting somewhat of a repulsion, perhaps to avoid protein–protein interactions. A portion of the effector region in rapamycin also shifts upon complexation. The triene region is fully conjugated and planar in the binary structure, but in the FRAP complex this region undergoes rotation about C18–C19 and C20–C21 which allows for the C23 methyl group to be buried in a small cleft in FRAP.

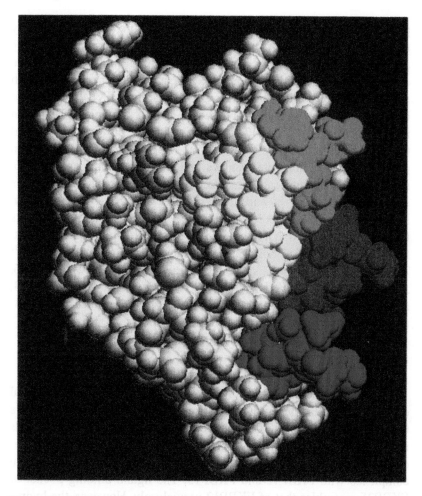

Figure 7. The composite surface of the FKBP12-FK506 complex which interacts with calcineurin is composed of the residues Asp-37–Asp-41 (yellow), Arg-42–Phe-46 (green), and His-87–Ile-90 (red) of FKBP12, together with the effector domain of FK506 (blue).

This longer loop region enables FRAP to bind to rapamycin and not its similar analogue, FK506.

FKBP13

The X-ray structure of the complex of FKBP13 with FK506 has been reported.[198] FKBP13 forms a tight complex with FK506, but this complex interacts weakly with calcineurin. The structure of FKBP13 is very similar to FKBP12 with identical binding pocket residues and 43% amino acid residue identity. The poorer CN inhibition is due to subtle changes in the protein's loop regions and surrounding residues which make up the composite surface for CN recognition. FKBP13 bound to FK506 consists of a four stranded β-sheet and a single α-helix. An interesting feature in FKBP13 is a disulfide bridge, between Cys-20 and Cys-75, in lieu of the loop crossing found in FKBP12 (vide supra). An overlay of the binding site residues and immunosuppressant, FK506, show nearly superimposable images. As discussed above, mutagenesis studies identified Gly-89 and Ile-90 in FKBP12 (Pro-97 and Lys-98 in FKBP13) as key residues for CN binding. Lys-98 in FKBP13 projects further into the hydrophobic binding cavity, thus disrupting the pocket. In addition, Arg-42 in FKBP12 is substituted by a Gln-50 in FKBP13.[47] This difference breaks the salt triad interaction found between Tyr-26, Asp-37, and Arg-42 in FKBP12 (Tyr-34, Asp-45, and Asn-50 in FKBP13, respectively). Substitutions in the 40s and 80s loop residues most probably account for the decrease in inhibition of CN.

FKBP25

The X-ray structure of the complex of rapamycin and the C-terminal domain of FKBP25 has been recently reported by Schreiber et al.[199] The C-terminal domain of FKBP25 is a peptidyl–prolyl isomerase and binds most strongly to the immunosuppressant rapamycin. The overall three-dimensional structure is similar to both those reported for the rapamycin complexes with FKBP12 and FKBP13. The ligand-binding pocket of FKBP25 resembles that of FKBP12 very closely. However, the hydrophobic pocket differs in surface area due to a F46L (Leu-162 in FKBP25) substitution. The smaller leucine residue causes the pocket to be less packed in van der Waals surface area. Other differences which may also account for the reversal in immunosuppressant binding preference are discussed below.

In FKBP25, the 40s loop contains eight more residues which includes a KKKK motif indicative of a nuclear localization signal site. This region is disordered in FKBP25, but is ordered in the FKBP12–rapamycin complex. Located in this disordered loop region is Asn-158 which corresponds to Arg-42 in FKBP12. The salt triad no longer can exist as can be seen from the displaced location of Asn-158 (Figure 8) In addition, due to the increased flexibility of the 40s loop in FKBP25, the distance

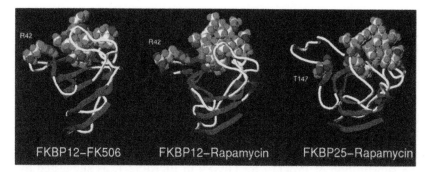

Figure 8. Ribbon diagrams of the FKBP12–FK506 complex (*left*), the FKBP12–rapamycin complex (*middle*), and the FKBP25–rapamycin complex (*right*). Arg-42 in FKBP12, and the corresponding residue Thr-147 in FKBP25, are shown in space-filling representation, as are the respective ligands.

between the hydrogen bond of Asp-146 (corresponding to Asp-37 in FKBP12) and the hemiketal group in rapamycin is less ideal and appears weakened in the complex structure.

In the region corresponding to the 50s loop in FKBP12, additional interactions between the protein and rapamycin are found. The E-54 substitution to a Lys-170 creates two additional hydrogen bonds to the ligand. Upon docking FK506 into the binding pocket, no interaction is seen between Lys-170 and FK506.

The region corresponding to the 80s loop in FKBP12 contains some significant changes also. First, Ala-206 in place of Ile-90 (FKBP12) creates a van der Waals contact between the alanine and rapamycin's C11 methyl residue. C11 has previously been identified as a residue which contributes to ligand binding. Second, Gln-203 in place of His-87 disrupts the interaction between the pyranose ring and the histidine residue. Taken together, the sum of these variations and substitutions must account for the increased binding preference for rapamaycin.

FKBP52

The solution structure of the N-terminal domain of FKBP52, which contains the first FKBP domain, has recently been solved using homo- and heteronuclear multidimensional NMR.[200] The N-terminal domain has 49% sequence identity with hFKBP12 and contains completely conserved binding pocket residues. The overall structure is very similar

to hFKBP12 and consists of six antiparallel β-sheets and one hydrophobic α-helix (see Figure 5). The uncommon loop crossing found in
FKBP12 is present also in FKBP52. Two hydrogen bonds between main
chain atoms in the loops ((98)NH–O(41), (41)NH–O(98)) help stabilize
the structure and make this crossing possible. Although the overall
solution structures appear to be very similar, the immunosuppressant
binding constant is roughly 2 orders of magnitude smaller for FKBP52-1.
In addition, the FKBP52-FK506 complex does not significantly inhibit
calcineurin. One possible explanation for the lower affinity for FK506 is
the substitution of Pro-119 for Gly-89 in FKBP52-1, as suggested by
mutation studies on FKBP12.

Comparison of FKBP Domains

Figure 5 compares the FKBP domains of FKBP25 (from the X-ray
structure of the complex with rapamycin) and FKBP52 (from the NMR
solution structure of domain 1) to FKBP12 (from the complex with
FK506). The R42T, E54K, and H87Q changes result in striking reversals
of the electrostatic environment surrounding the pocket. These changes
are visualized in Figure 5 in which the electrostatic potential has been
mapped onto the Connolly surface of the two domains. The electroposi-
tive regions of FKBP12 are electronegative (blue coloring) in FKBP25,
while the negatively charged Glu-54 of FKBP12 has been replaced by
the positive Lys-170 in FKBP25. Thus, while the pocket itself is largely
unchanged, the environment around the surface of the cavity is quite
different electrostatically. The FKBP domain may thus represent a common geometric recognition module, perhaps binding proline-containing
reverse-turn motifs, while the alterations around the pocket confer dif-
fering specificities.

B. Cyclophilins

Cyclophilin A

Cyclophilin A, an antiparallel β-barrel formed by eight β-strands
connected at the top and bottom by loops and α-helices (Figure 4), is an
approximately spherical molecule 34 Å in diameter.[201] Cyclophilin's
topology (+1, –3, –1, –2, +1, –2, –3) is unique, as is FKBP12's. Unlike
β-barrels found in some transport proteins, cyclophilin's core is filled
with hydrophobic residues which preclude ligand binding in the barrel
center.[202]

As in the case of FKBP12, analysis of complexes of cyclophilin A with various ligands have allowed delineation of the cyclophilin rotamase domain. The first reported three-dimensional structure of cyclophilin was as a complex with a tetrapeptide substrate, *N*-acetyl-Ala-Ala-Pro-Ala-amidomethylcoumarin, solved using a combination of NMR (solution) and X-ray (crystal).[75] Additional complexes with dipeptides have been solved,[74,77-79] as well as the CsA–CyP complex.[203-208] In the CsA–CyP complex, only part of cyclosporin A makes contact with the protein, analogous to the FKBP–FK506 case (Figure 4). CsA residues 9, 10, 11, 1, 2, and 3 (Figure 2) are in direct contact with the protein, with the remaining residues extending into the solvent accessible region of the protein face. The CyP binding domain, like that of FKBP12, comprises 13 residues: Arg-55, Phe-60, Met-61, Gln-63, Gly-72, Ala-101, Asn-102, Ala-103, Gln-111, Phe-113, Trp-121, Leu-122, and His-126. The proline binding pocket is formed by Phe-60, Met-61, Phe-113, Trp-121, Leu-122, and His-126. Figure 9 shows the residues that comprise the CyP domain, and the solvent accessible surface of the binding domain.

CyP-domain mutants of CyPA have separated rotamase activity, CsA binding, and calcineurin binding. Replacement of Trp-121 with Ala reduced enzymatic activity by an order of magnitude but decreased CsA binding 200-fold,[209,210] demonstrating that PPIase activity and CsA binding can be decoupled. Other mutations in the rotamase domain of CyPA studied include H54Q, R55A, F60A, Q111A, F113A, and H112Q.[211] The changes causing the greatest loss of PPIases activity were in residues R55, F60, and H112, suggesting they may be involved in the enzymatic mechanism (Figure 9). The PPIase-inactive mutants were capable of inhibiting calcineurin in the presence of CsA. Other changes affect calcineurin inhibition but not rotamase activity: changing R69, K125, and R148 to either negatively charged or uncharged residues has little effect on enzymatic activity but greatly reduces calcineurin inhibition.[96]

Cyclophilin B

The X-ray structure of the complex of the D-Ser cholinyl ester of CsA and human recombinant CyP-B has been reported.[212] Figure 10 compares this structure with that of CsA bound to CyP-A. The proteins are depicted in ribbon form. Overall, the structures of the two cyclophilins are closely similar. Both the N- and C-termini of CyP-B are extended relative to CyP-A. The N-terminal β-strand of CyP-B bends toward the β-barrel;

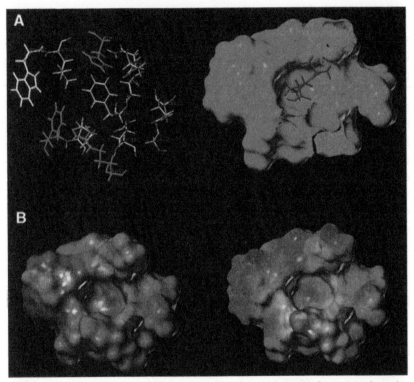

Figure 9. (**A**) (*left*) Residues comprising the rotamase domain of cyclophilin A. Residues that appear to be involved in the mechanism of
enzymatic activity are colored green; Trp-121 (yellow) is important for
cyclosporin A binding. (*right*) The solvent accessible surface of the cyclophilin A domain, from the complex with dipeptide Ala-Pro. (**B**) Lipophilic
(*left*) and electrostatic (*right*) potential surfaces of the cyclophilin A domain.

the C-terminal strand ends in a β-turn and a final short β-strand (Figure
10). Two of the loops surrounding the rotamase domain differ between
the two structures. The short loop connecting the first two β-strands in
CyPB (Ile-20–Gly-21–Asp-22–Glu-23, corresponding to Val-12–Asp-
13–Gly-14–Glu-15 in CyPA) is canted at a different angle than in CyPA,
and forms a type II' β-turn rather than the type I' turn found in CyPA. A
more significant difference is found in the loop between the second
a-helix and following β-strand. This loop (residues Thr-153–Lys-163,
corresponding to Phe-145–Lys154 of CyPA) adopts a quite different
conformation. The rotamase domain itself has the same structure in the
two cyclophilins.

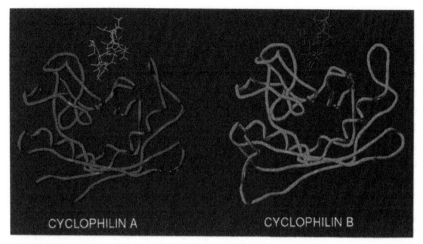

Figure 10. Ribbon diagrams of cylophilin A and cyclophilin B complexed with cyclosporin A.

Cyclophilin C

The complex of cyclophilin C from mouse kidney with cyclosporin A has also been solved by X-ray crystallography.[213] mCyPC has the same β-barrel structure as hCyPA. The rotamase domain is conserved, and the binding of CsA to mCyPC is essentially the same as to hCyPA. Significant differences are observed in three loop regions, however: Gln-179–Thr-189, Asp-47–Lys-49, and Met-170–Ile-176. The Cα atoms of all three loops show significant migration compared to CyPA, and the first two loops have different backbone conformations. It was noted above that CyPC, but not CyPA, binds a 77 kDa glycoprotein called CyCAP, and this interaction is competitively inhibited by CsA, suggesting that CyCAP and CsA may bind at the same or overlapping domains. Of the three loops mentioned, only the Gln-179–Thr-189 loop is close to the binding domain, and the positional displacements and conformational differences noted above may play a role in CyCAP binding.

C. Structural Analysis of Protein–Ligand Complexes in Drug Design: Examples from Immunophilin Research

The combination of rapid advances in the fields of molecular biology, protein structure determination, and molecular modeling has resulted in the emergence of structure-based drug design as a widely used and powerful tool in pharmaceutical research.[214-217] Structural information for target proteins and their complexes with tight-binding inhibitors or ligands is typically obtained by X-ray crystallography or high-field, multidimensional NMR.[218-220] These two techniques are in many respects complementary to each other. While both determine the atomic coordinates of molecular ensembles, X-ray crystallography is suitable for quite large molecules ($M_r > 10,000$) and the data represent time averages on the scale of seconds to hours, while NMR is limited to considerably smaller molecules and the time scales are on the order of nanoseconds to seconds. Although the proteins that can be studied by NMR are limited in size, the time scales accessible by this method make it extremely useful for studying dynamic processes in macromolecules. We will discuss the use of both X-ray crystallography and NMR spectroscopy in the study of the binding of small ligands to immunophilins.

The "effector domain" hypothesis of the mechanism of action of FK506, rapamycin, and cyclosporin A was discussed in an earlier section. An extensive body of work was developed by various research groups in the pharmaceutical industry seeking to develop analogues of the immunosuppressant drugs which were simpler structurally and had improved safety profiles. Holt's group at SmithKline Beecham was the first to publish on the SAR of simple ligands for FK506, based on their formulation of the minimal binding domain (Figure 11).[188,221] Their work established that simple acyclic N-glyoxyl pipecolyl esters retain good affinity for FKBP12 and are potent inhibitors of its enzymatic activity. Several compounds described by these workers are shown in Figure 11 (1–6). Compound 1, a simple ester of pipecolic acid with a cyclohexyl glyoxyl appendage to mimic the pyranose region of FK506, has a K_i of 2 µM. Replacing the cyclohexyl group with the tertiary dimethylpropyl moiety (2) resulted in a three-fold increase in affinity for FKBP12; further addition of the more hydrophobic phenylpropyl ester side chain produced a simple ligand with submicromolar affinity (3, $K_i = 110$ nM). Substituting the aryl ester with the 3,4,5-trimethoxyphenyl moiety, or adding a second aromatic ring α to the ester oxygen, produced extremely high-affinity FKBP ligands (4 and 5, respectively).

The 3,4,5-trimethoxyphenyl group is also an excellent replacement for the pyran moiety of FK506, as reported by workers at Vertex Pharmaceuticals.[189] Compound 7 is a very potent FKBP inhibitor, and the branched ester "semaphore" compound 8 is equipotent with FK506 as an FKBP12 inhibitor and ligand. Replacement of the pipecolic acid ring

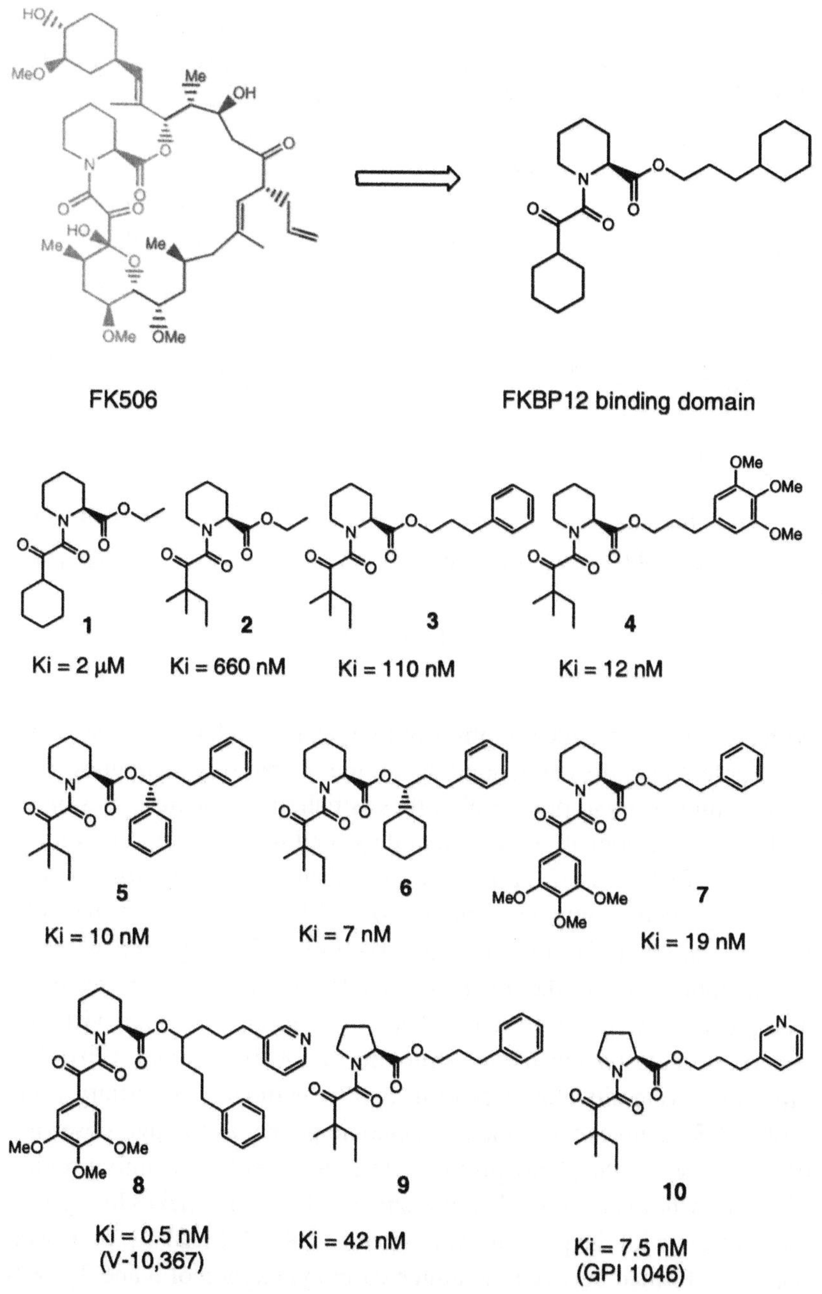

FK506

FKBP12 binding domain

1

Ki = 2 µM

2

Ki = 660 nM

3

Ki = 110 nM

4

Ki = 12 nM

5

Ki = 10 nM

6

Ki = 7 nM

7

Ki = 19 nM

8

Ki = 0.5 nM
(V-10,367)

9

Ki = 42 nM

10

Ki = 7.5 nM
(GPI 1046)

Figure 11. Binding domains of FK506 together with representative small molecule mimetics of the FKBP-binding domain. The latter is shown in blue; the effector domain is in red.

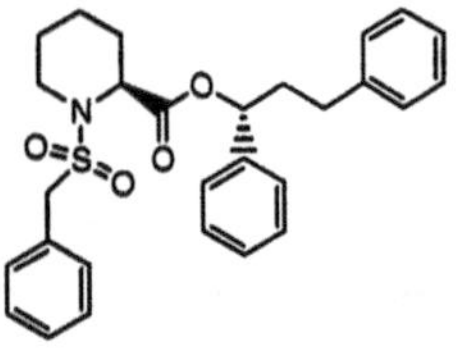

11

Ki = 160 nM

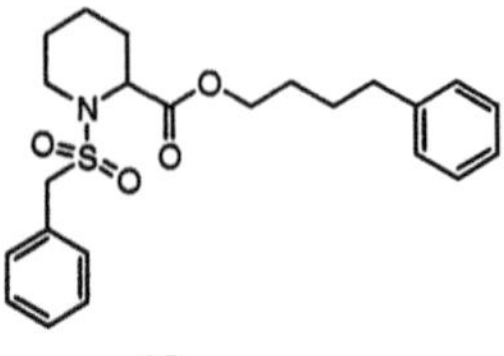

12

Ki = 34 nM

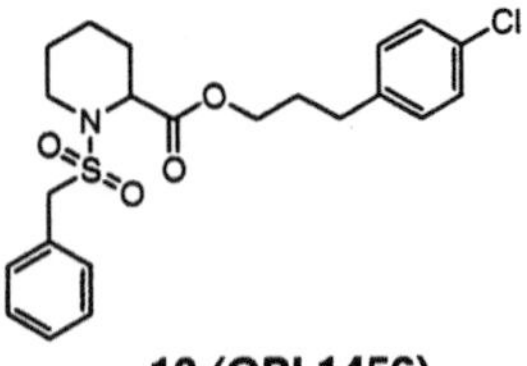

13 (GPI 1456)

Ki = 1705 nM

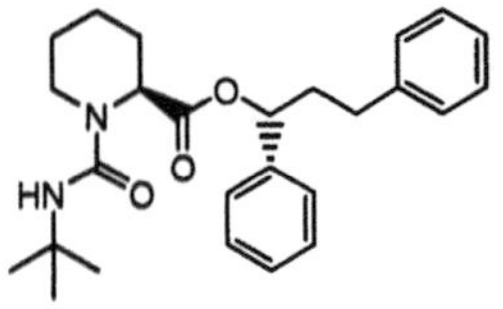

14

Ki = 350 nM

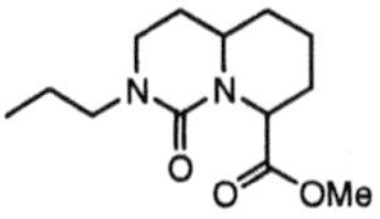

15

Ki = 2400 nM

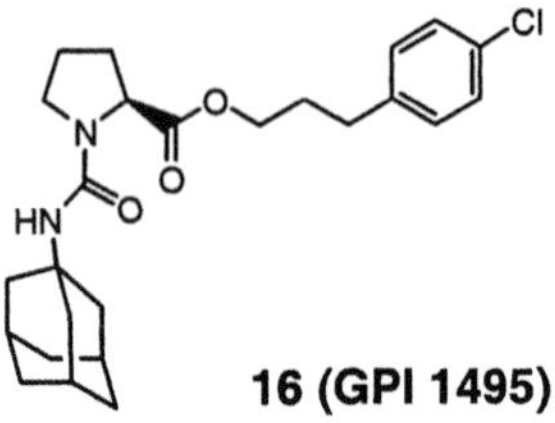

16 (GPI 1495)

Ki = 2340 nM

Figure 11. Continued.

with proline also produces high-affinity ligands which in some cases are more potent than the corresponding six-membered ring analogues.[222]

Structural analysis of several of these small molecule mimetics of the FKBP-binding domain indicate that they bind in a manner very similar to that of FK506 itself. Holt et al. described the crystal structures of several synthetic compounds bound to FKBP12, including **5** and **6**.[188] Figure 12a, taken from the crystal structure of **5** complexed with the immunophilin, shows the ligand surrounded by the active site residues. The pipecolinic acid ring and α-ketoamide groups of **5** are virtually superimposable on the bound structure of FK506 (Figure 12b). The pipecolinic ring sits atop Trp-59 in both structures. The tertiary alkyl group of **5** occupies the same hydrophobic pocket as the pyranose ring of FK506, while the phenylpropyl ester group occupies a similar region of space as the cyclohexylethenyl group of FK506, and makes hydrophobic contacts with the protein. Analogous to FK506-FKBP12, hydrogen bonds are formed between the amide carbonyl oxygen of **5** and Tyr-82-OH, and the ester carbonyl oxygen of the ligand and Ile-56-NH.

Similar results were reported by Vertex for **8** bound to FKBP12.[189] While the pipecolinyl and dicarbonyl regions bind almost identically to

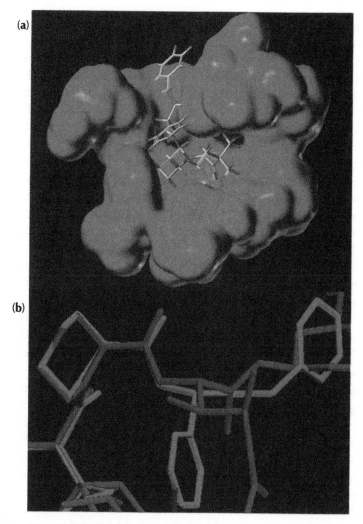

Figure 12. (a) FKBP ligand **5** bound to the rotamase domain of FK506. The solvent-accessible surface of the protein is in green. (b) Overlay of the bound conformations of **5** (green) and FK506 (red).

FK506, the trimethoxyphenyl group is in partial conjugation with the keto carbonyl of the dicarbonyl moiety and is not in the plane of the pyranose group of FK506. Some distortion of the 80s loop is caused by the bulky substituted aryl group. One of the phenyl rings of the flexible branched ester chain binds similarly to the cyclohexyl group of the immunosuppressant; the other lies over Phe-46 and Tyr-26.

Based upon consideration of these structures and molecular modeling, workers at Guilford Pharmaceuticals evaluated proline-based structures such as **9** and **10**. It was anticipated that the proline ring would be capable of sinking somewhat deeper into the binding pocket of FKBP12 and binding more tightly. Compounds **9** and **10** are high-affinity ligands for FKBP. Lamb and Jorgensen have recently reported the results of a study of the binding affinities of several FKBP12 ligands, including **3**, **5**, **9**, and **10**, using free-energy perturbation techniques in Monte Carlo statistical mechanics simulations.[223] The computational studies of these investigators predicted that the prolyl compounds reached further into the pocket and bound more favorably. The tighter binding predicted for the prolyl structures was attributed to the removal of an unfavorable van der Waals interaction with Trp-59, relative to the pipecolinyl ring, and increased interaction with Tyr-26.

Replacement of the dicarbonyl functionality of compounds such as **5** with other chemical groups, such as sulfonamide (**11**) and urea (**16**) has been described. Compound **11**, in which the dimethylpropyl glyoxyl moiety of **5** has been replaced by benzylsulfonyl, has a K_i of 160 nM. Similar compounds (e.g. **11** and **13**) have been reported by the Guilford group.[222] No detailed structural information is available to date on the complexes of these compounds with FKBP12.

Urea linkages have also been explored as replacement for the dicarbonyl group. The Agouron group has described the crystal structures of several of these compounds (e.g. **14** and **15**) with FKBP12. In the complex between **14** and FKBP12, the pipecolinyl ring and ester side chain were nearly identical to the corresponding regions of ligand **5** in its complex with the protein. Hydrogen bonds are formed between the urea carbonyl and Tyr-82-OH, and the urea –NH and the Asp-37 carboxylate.

NMR spectroscopy is an extremely valuable tool for studying the dynamic and kinetic properties of proteins, as well as their three-dimensional structures. NMR techniques have evolved rapidly in the past two decades and have been used extensively to characterize the immunophilins and their ligand complexes. Protein NMR has become a tool as valuable as X-ray crystallography in the practice of structure-based drug design. As discussed previously, the study of the immunophilins has been a fertile area for exploring new ways of integrating chemistry and biology. It has likewise been a fruitful arena for the application of modern spectral techniques to the study of the relationship between biological activity and protein structure and dynamics. We will discuss several examples of the application of modern NMR methods in structure-based drug design as exemplified in immunophilin research.

Perhaps the most perplexing problem in three-dimensional solution structure determination by NMR is the complex signal overlap. As molecules increase in size, the one-dimensional ^{1}H spectrum becomes more complex and it becomes difficult to assign corresponding peak shifts due to the large increase in the number of resonances. In addition, large proteins exhibit slow tumbling motions in solution, resulting in line-width broadening. Recent advances in the area of NMR spectroscopy have led to improvements in overall spectral resolution and sensitivity through the development of multidimensional pulse sequences.[224,225] Through the biosynthetic incorporation of stable isotopes (^{15}N, ^{13}C) in proteins, spectra are simplified and assignment of residue peak shifts is facilitated. Distance constraints obtained from analysis of multidimensional spectra are used with molecular modeling software to generate families of energetically favorable conformations based on experimental data.

NMR can be used for directly identifying molecules which bind to proteins in a mixture of binders and nonbinders. Fesik and coworkers demonstrated this technique using relaxation edited experiments on a mixture of several low molecular weight compounds with macromolecules.[226] Signals from the protein and bound ligands are eliminated, thereby enabling only those resonances which bind to be identified from the difference spectrum. Fesik's group demonstrated the utility of this method for screening compound mixtures to discover new protein ligands, applying it to the discovery of new inhibitors of FKBP12[227] and stromelysin.[228]

Backbone amide shift peaks can be systematically monitored via 2-D HSQC experiments during titration studies. Shown in Figure 13 are ^{15}N FKBP12 titrations with two immunosuppressants (FK506 and rapamycin) and three non-immunosuppressant FKBP12 ligands representing three different classes (**10** [GPI 1046], **13** [GPI 1456], and **16** [GPI 1495], containing α-ketoamide, sulfonamide, and urea moieties, respectively; Figure 11). Similar residues which shift due to ligand binding enable identification of those residues which are involved in the binding pocket or neighboring regions affected upon complex formation. In Figure 13, titration spectra of FKBP12 with FK506, GPI 1046, and rapamycin show significant shifts in similar peak resonances. The numbered residues are those which are involved in the ligand-binding pocket or surrounding loop regions, or those which exhibit sizeable shift changes. Both FK506 and rapamycin bind in similar fashion as previously discussed. Most interesting in comparing these three titration studies is perhaps the "mode" in which binding seems to be occurring. Both FK506 and rapamycin seem to bind with a slow exchange rate as noted from the sudden increase in shift changes at higher ligand concentrations. GPI 1046 shows a more gradual change in peak shifts as the concentration of ligand approaches saturation. Figure 13c is an overlay of the saturation point of all three ligands above. It can be seen that all three ligands effect the complex conformation in a somewhat different fashion.

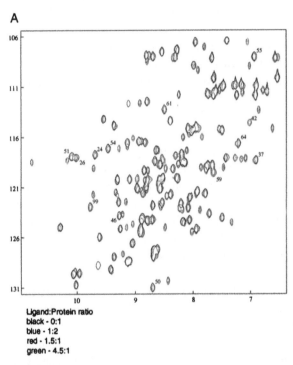

A

Ligand:Protein ratio
black - 0:1
blue - 1:2
red - 1.5:1
green - 4.5:1

Figure 13. FKBP12 titration spectra with FK506, rapamycin, and nonimmunosuppressive ligands, GPI 1046, 1456, and 1495. All titration data were collected at 30 °C on a 600 MHz GE NMR in PBS buffer, 5 mM DTT, 2 mM NaN3, 10% D2O, pH 7.4, and 15 N FKBP protein concentration of approximately 1 mM. Ligands were either dissolved in PBS or in combination with ethanol to increase solubility. (**A**) FK506 titration. (**B**) GPI 1046 titration. (**C**) Rapamycin titration. (**D**) Overlay of saturation endpoints for FK506, GPI 1046, and rapamycin. (**E**) GPI 1456 and 1495 overlay. (**F**) Overlay of all three nonimmunosuppressive ligands at saturation point.

B

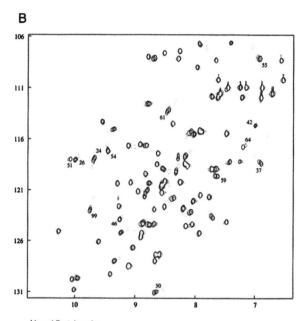

Ligand:Protein ratios
black - 0:1; blue - 1:2; red - 2:1; green - 3:1; purple - >5:1

C

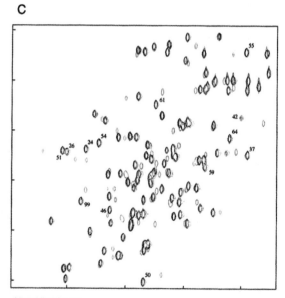

Ligand:Protein ratio
black - 0:1
blue- 1:2
red - 2.5:1
green - 3.5:1

Figure 13. Continued.

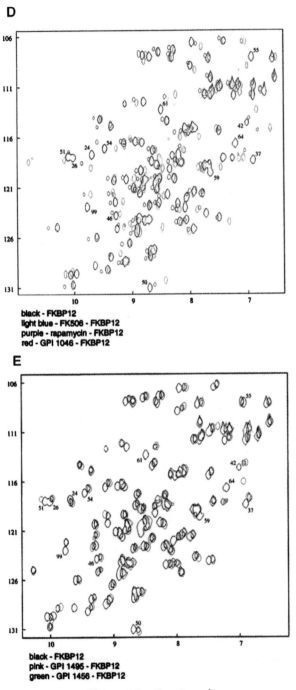

D

black - FKBP12
light blue - FK506 - FKBP12
purple - rapamycin - FKBP12
red - GPI 1046 - FKBP12

E

black - FKBP12
pink - GPI 1495 - FKBP12
green - GPI 1456 - FKBP12

Figure 13. Continued.

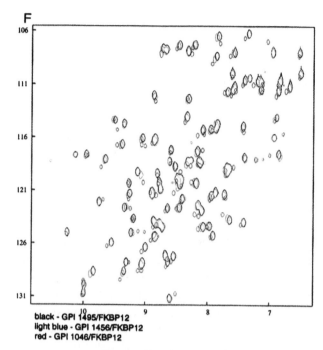

black - GPI 1495/FKBP12
light blue - GPI 1456/FKBP12
red - GPI 1046/FKBP12

Figure 13. Continued.

Figure 13d is an overlay of the saturation point of GPI ligands, 1495 and 1456, urea, and sulfonamide ligands, respectively. Almost exactly superimposable spectra suggest that the binding and conformation of the two complexes are closely similar. Due to the different structural classes of the ligands, one might expect to see significant differences in the conformation of the complexes. Therefore, this overlay suggests very similar effects on the conformation of the complex upon binding and complete three-dimensional structure determination would give insight into the exact structural models of the complex.

The last panel of Figure 13 shows all three GPI ligands at the saturation point overlaid. Here again, GPI 1046 seems to be binding in a slightly different manner. However, from analysis of the peaks which shift due to binding interaction, one can obtain a general starting point for studying protein–ligand interactions.

Rational drug design is dependent upon understanding not only the molecular structure of the drug target, but also the molecular motions. Early studies of bound and unbound FKBP12 concluded that the protein was little affected by ligand binding, its conformation in complex with

FK506 or rapamycin being "nearly identical" to the unbound state. Differences observed in the three structures were ascribed to experimental artifacts such as crystal-packing interactions and lack of NMR restraints. However, more sophisticated studies utilizing multidimensional and isotope-edited NMR techniques suggest that significant local modulation of conformational flexibility of FKBP12 may result as a consequence of ligand binding, and may have functional implications.

Backbone dynamics of FKBP12 were analyzed using ^{15}N NMR relaxation experiments[229] and molecular dynamics simulations.[186] Such experiments identify regions of flexibility which are most likely to undergo structural and/or dynamical changes upon binding. While most of FKBP12 is highly ordered in solution, several residues in the 36–45 and 78–95 loops were observed to have significantly greater mobility than the rest of the molecule. Molecular dynamics simulations of the uncomplexed protein suggest that the 85–94 loop region may move as a hinged flap capable of covering part of the binding domain. Both NMR dynamics studies and molecular dynamics simulations show that this 80s loop region becomes fixed upon binding FK506; in contrast, binding of rapamycin causes considerably less loss of mobility in the flap, due to the fewer van der Waals interactions made by the pyran ring of rapamycin compared to FK506.

X-ray and NMR studies have identified two different conformers of FK506 bound to wt FKBP12 and the R42K, H87V double mutant of FKBP12.[230,231] The complex of the latter with FK506 is ineffective as an immunosuppressant. Lepre et al. showed that the dmFKBP12–FK506 complex exhibits structural differences in regions outside of the binding pocket, i.e. solvent-exposed regions of the ligand. The most significant changes were in the macrocyclic backbone near C16. This region lies close to Arg-42 in wtFKBP12 and the Arg-42 to Lys-42 side-chain mutation affects the overall conformation. Interestingly, the complexes of FK506 with the R42K and H87V single mutants show the same conformation as the wild type complex; only the double mutation induces significant local changes in ligand conformation. Lepre et al. noted also that replacement of Arg-42 with lysine replaces the Arg-42-Asp-37 salt bridge of wtFKBP12 with a hydrogen bond, and results in Lys-42 having a more extended conformation than Arg-42 in the native protein. This extended conformation displaces the 15-OMe in FK506, which results in internal strain which can be relieved by changing the ligand backbone conformation. The resulting induced 90° shift in the position of 13-OMe is more readily accommodated when His-87 is replaced by the sterically less demanding Val.

These studies provide a fascinating example of the subtle interplay between local protein conformation and ligand conformation as the two partners adjust to find the most favorable complex. Molecular dynamics simulations of FK506 binding to native FKBP12 suggest the dynamical interplay between protein and ligand during the binding process.[232] Ivery and Weiler's computational modeling studies suggested that initial binding of the pipecolate region of FK506 and the hydrophobic pocket of FKBP12 is followed by a hingelike movement of the 80s loop to close over the pyranose region of the ligand. This induces conformational shifts in the 13- and 15-methoxyl groups to assume their observed bound conformations. The complex is further sealed by movement of the 40s loop.

Three-dimensional structures of ligand–receptor complexes can be calculated without knowledge of the receptor structure itself, as long as the ligand pocket is known. Such calculations were done on FK506 bound to triple-mutant FKBP13.[233] tmFKBP13–FK506 inhibits calcineurin as potently as FK506–FKBP12, whereas wild-type FKBP13–FK506 exhibits significantly less inhibitory potency.[234] ^{13}C-labelled FK506 bound to tmFKBP13 shows exceptional similarity in overall three-dimensional structure with FK506 bound to wtFKBP12. Time-averaged restrained molecular dynamics simulations showed a convergence to a single family of minimum energy structures. The data suggest that the mutations (Q50R, A95H, K98I—corresponding to R42, H87, and I90 in FKBP12, respectively) affect calcineurin interaction rather than ligand conformation.[235]

V. PROLYL ISOMERASES IN THE NERVOUS SYSTEM[236]

A. Neuroimmunophilins and Nervous System Effects of Immunosuppressant Drugs

Soon after the discovery of FKBP12, Northern blot analysis of mRNA demonstrated the presence of the immunophilin throughout the body, including the brain.[237] When rat brain homogenates were monitored for [³H]FK506 binding, Snyder's group found that FKBP12 levels in the brain were much higher than in immune tissues, nearly 50-fold greater.[238] This discovery was made just at the time of the literature reports on calcineurin as the target of the FKBP12–FK506 and cyclophilin–CsA complexes. Using autoradiography, immunohistochemistry, and in situ hybridization, the Hopkins group determined the brain localization of FKBP12, cyclophilin A, and calcineurin. Both FKBP12 and CyPA were present in high levels and primarily neuronal.[238,239] The regional distribution of FKBP12 mapped closely to that of calcineurin, a finding immediately suggestive of a physiological association. Cyclophilin's distribution also paralleled that of calcineurin extensively, although it was found in some regions lacking calcineurin. Particularly high levels of immunophilins are found in the hippocampus, dopaminergic neurons of the nigrostriatal pathway, and cerebellar granule cells. The distributions of FKBP12 and CyPA do not entirely overlap: CyPA is enriched in the brain stem and largely absent from the caudate nucleus, while the opposite is true of both FKBP12 and calcineurin.

Since the colocalizations of the immunophilins with calcineurin suggested a functional link, Snyder's group looked for proteins in brain homogenates whose phosphorylation levels were enhanced by incubation with FK506. One such protein was the enzyme nitric oxide synthase (NOS).[240] NOS generates nitric oxide (NO) from arginine, and NO, acting as a rapidly diffusing short-lived messenger molecule, plays numerous roles in the body.[241,242] Macrophages generate NO in response to endotoxin; in blood vessels, NO is a regulator of vasodilation; and in the brain, it appears to act as a gaseous neurotransmitter. Neuronal NOS is rendered inactive by phosphorylation with protein kinase C,[243] and thus calcineurin regulates NOS activity in a calcium-dependent manner by dephosphorylating it. One pathway for calcium-mediated activation of NOS involves the N-methyl-D-aspartate (NMDA) subtype of glutamate receptor. Stimulation of NMDA receptors by the excitatory neurotransmitter glutamate results in opening of voltage-dependent calcium channels and influx of calcium, resulting in activation of NOS and subsequent cGMP formation.[244] The well-known neurotoxic properties of excessive levels of glutamate are thought to be mediated at least in part by neuronal NOS. Glutamate and NMDA are toxic to neuronal cultures, and this toxicity is blocked by NMDA antagonists; NMDA antagonists also block the neuronal damage associated with vascular stroke, suggesting a pathological role for glutamate in ischemic brain damage.[245] Glutamate neurotoxicity is also blocked by NOS enzyme inhibitors,[246] and in NOS-deficient mice.[247] Since inhibition of calcineurin by FK506 and CsA would result in deactivation of NOS, these drugs should be capable of exerting neuroprotective effects similar to NMDA antagonists. Experimental work demonstrated that this was indeed the case. In cortical neuronal cultures, both FK506 and CsA blocked NMDA neurotoxic-ity.[240] As would be expected for an effect mediated by FKBP12 and calcineurin, the neuroprotective effects of FK506 were antagonized by rapamycin. FK506 was shown by several groups to be effective in rodent models of focal and global ischemia.[248–250] In the rat middle cerebral artery occlusion (MCAO) model, FK506 reduced cortical damage when given one-hour postocclusion, and its protective effects were blocked by rapamycin.[248] The protective effects of FK506 are not necessarily en-tirely or primarily due to protection against glutamatergic neurotoxicity, however. Butcher et al. have reported results in the focal ischemia model that are inconsistent with inhibiting glutamate toxicity as a protective mechanism,[251] and cyclosporin A reportedly fails to protect against transient global ischemia.[252] It is entirely possible that at least some of the neuroprotective effects of FK506 occur through a non-calcineurin pathway.

FK506 can effect glutamatergic neurotransmission by another path-way: modulating neurotransmitter release. NMDA-stimulated release of glutamate from brain synaptosomes is inhibited by FK506, and depolarization-induced neurotransmitter release from PC12 cells is also prevented by FK506.[253] These effects are antagonized by rapamycin. However, depolarization-induced neurotransmitter release from synap-tosomes is enhanced by low levels of CsA as well as an FK506 analogue (FK520)![253] Glutamate release from synaptosomes treated with potas-sium channel blockers is also potentiated by immunophilin ligands.[254] The varying effects of the immunophilin ligands on neurotransmitter release may be a function of which specific calcineurin substrates are affected. Potential players include proteins such as synapsin I and dynamin I, which promote neurotransmitter release when phosphory-lated.[253,254] Increased phosphorylation resulting from calcineurin inhibi-tion might result in enhanced transmitter release. On the other hand, NO is required for neurotransmitter release from NMDA-stimulated synap-tosomes,[255] so in this context calcineurin inhibition would inhibit release.

Another protein whose phosphorylation was augmented in the pres-ence of FK506 was growth-associated protein-43 (GAP-43),[238] which mediates neuronal process extension.[256] Several other pieces of data were suggestive of a functional link between FKBP12, GAP-43, and nerve growth or regeneration. Both proteins are upregulated following damage to facial or sciatic nerves when regenerative processes are occurring.[257] The growth cones of neonatal neurons are particularly enriched in FKBP12 (as are peripheral nerves).[238] The association of GAP-43 and

FK506 led the Hopkins group to investigate the possible effects of FK506 on nerve growth.

Pheochromocytoma, or PC-12, cells are a neural crest-derived cell line which respond to nerve growth factor (NGF) by differentiating into a neuron-like phenotype and extending neurites. NGF promotes neurite extension in these cells with an EC_{50} of approximately 100 ng/mL. Surprisingly, FK506 potently potentiated the effects of NGF in this cell system in a dose-dependent manner.[258] Remarkably, the EC_{50} for FK506 to promote neurite outgrowth in the presence of a submaximal (1 ng/mL) concentration of NGF was found to be 500 picomolar. In the absence of exogenous NGF, FK506 was ineffective in PC-12 cells. This result suggested that the compound worked by enhancing the sensitivity of the PC-12 cells to the trophic factor. Rapamycin and cyclosporin A produced similar results in PC-12 cells. The result with rapamycin was extremely significant. In the NOS-related studies described above, rapamycin consistently antagonized the effects of FK506, consistent with a calcineurin-mediated mechanism. The nerve growth experiments suggested that a non-calcineurin-mediated pathway was operative. Thus, although the initial impetus for this investigation was the effect of FK506 on GAP-43 phosphorylation via calcineurin inhibition, the positive activity of rapamycin suggested that these effects may not be relevant to the nerve growth properties of FK506.

The compounds were also active in cultured sensory neurons from chick dorsal root ganglia (DRGs).[258] Scoring the actions of the drugs by counting the number of processes longer than the DRG explant diameter 48 hours after drug treatment, it was found that FK506 and rapamycin promoted process extension with half-maximal effects produced at 500 pM concentration (Table 3). The immunophilin ligands produced neurotrophic effects in the neuronal cultures in the absence of added NGF, but it is possible that the drugs work by potentiating the effects of NGF produced by Schwann cells in the primary cultures; the effects of FK506 are partially (though not entirely) blocked by an antibody to NGF.[236b]

Similar in vitro neurotrophic effects of FK506 were reported using SH-SY5Y cells.[259] Growth factors such as NGF are known to produce inverted or bell-shaped dose response curves with respect to functional responses, and it is interesting that two different reports note that FK506 actually inhibits nerve growth in SH-SY5Y cells[259] and neuronal cultures at concentrations of 1 and 50 μM, respectively.[260]

Table 3. In Vitro and In Vivo Neurotrophic Effects of Immunosuppressant Drugs and Their Nonimmunosuppressive Analogues

Compound	K_i, nM	ED_{50}, nM	
FK506	0.4	0.5	
L-685,818	0.7	5.0	
Rapamycin	0.2	0.5	
WAY-124,466	12.0	25.0	
Cyclosporin A	20.0	50.0	
6-[Me]-Ala-CsA	50.0	50.0	
Treatment	Axonal dia. (mm)	Cross sectional area (mm^2)	Myelination
Sham	2.98 ± 0.221	9.14 ± 0.835	60.07
Lesion/Veh	1.56 ± 0.73	2.74 ± 0.357	5.30
Lesion/FK506	1.99 ± 0.192	4.16 ± 0.403	11.67
Lesion/L,685,818	1.87 ± 0.141	3.43 ± 0.295	16.68

These in vitro experimental findings were given clinical relevance in Gold's experiments on the effects of FK506 on regrowth of damaged peripheral nerves. In rats whose sciatic nerves were lesioned by crushing with forceps, daily treatment with FK506, administered subcutaneously, accelerated nerve regeneration and functional recovery.[261,262] Animals receiving FK506 regained use of the affected limb more rapidly than control animals, and this was reflected morphologically in the observation that drug-treated animals had larger axonal calibers and increased reinnervation of the muscle fibers. Also consistent with the in vitro data was the observed bell-shaped dose-response curve for FK506 in these studies.[259]

B. Separation of the Immunosuppressant and Neurotrophic Effects of FK506

The results with FK506, rapamycin, and cyclosporin A discussed above indicate that immunophilin ligands possess neuroprotective and neuroregenerative properties, and that both calcineurin-dependent and calcineurin-independent mechanisms are operative. The observation that both FK506 and rapamycin produced neurotrophic effects in vitro suggested that the nerve growth effects of the drugs were mediated, at least in part, by a non-calcineurin mechanism.

The use of nonimmunosuppressive analogues of FK506, rapamycin, and cyclosporin A to probe the mechanism of immunosuppressant action was discussed previously. These compounds also proved useful in exploring the neurotrophic actions of the immunosuppressants. Steiner et al. evaluated the in vitro and in vivo neurotrophic effects of three of the compounds in Figure 3, and compared them to the parent immunosuppressant drugs.[263] Both L-685,818 and WAY-124,466 (nonimmunosuppressive analogues of FK506 and rapamycin, respectively) were potent neurotrophic agents in PC12 cells and cultured chick dorsal root ganglia (Table 3). Both cyclosporin A and its 6-methyl analogue likewise produced neurotrophic effects in vitro, but in this case the nonimmunosuppressive compound was more potent and produced a greater maximal effect than its parent compound. Like their parent compounds, the nonimmunosuppressive analogues did not stimulate neurite outgrowth in PC12 cells in the absence of NGF; exogenous NGF was not required to produce effects in the DRGs. When compared to each other in a rat sciatic nerve crush model similar to that described by Gold, L-685,818 and FK506 produced comparable results. Animals receiving either drug had more large axons, and axons with a larger average diameter than vehicle-treated animals (Table 3). On day 14 of the experiment, all animals (vehicle- and drug-treated) showed the same ability to bear weight on the injured limb; by day 18, animals receiving either of the immunophilin ligands had nearly completely regained the ability to use the injured limb. The data for the immunosuppressive and nonimmunosuppressive compounds is summarized in Table 3.

The foregoing results demonstrated that the neurotrophic and immunosuppressant effects of FK506, rapamycin, and cyclosporin A could be separated, and that inhibition of calcineurin or FRAP is not involved in the nerve regeneration properties of the drugs. Hamilton et al. proposed that the nerve-regenerative effects of FK506 resided within the FKBP-binding domain.[264] According to this hypothesis, compounds which mimic the FKBP-binding portion of FK506, such as those depicted in Figure 11, should possess neurotrophic activity. A large number of such structures (including several previously reported in the literature) were synthesized and tested for neurotrophic activity (Figures 11 and 14).[222,263,264] Promotion of fiber extension (neurite outgrowth) from cultured chick sensory neurons was used as an in vitro assay for nerve growth effects. Compounds were also tested for their ability to inhibit FKBP12 rotamase activity. Table 4 presents this data for a number of N-glyoxyl prolyl and pipecolyl esters.

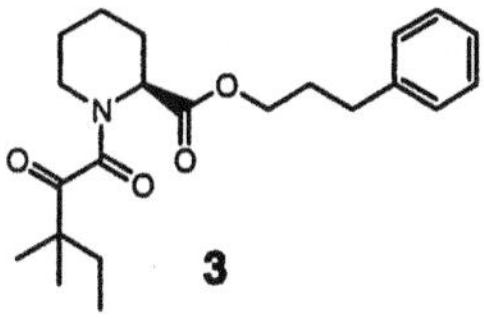

Figure 14. FKBP ligands that have been shown to promote regeneration of damaged nerves in vivo.

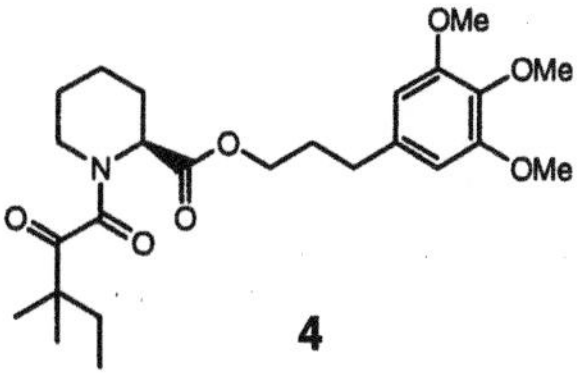

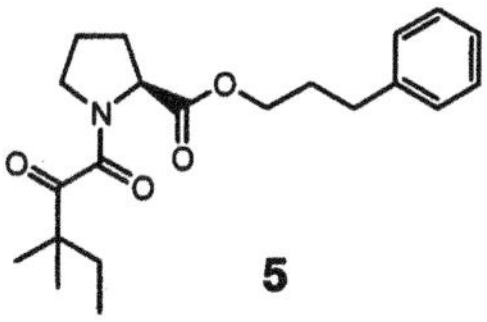

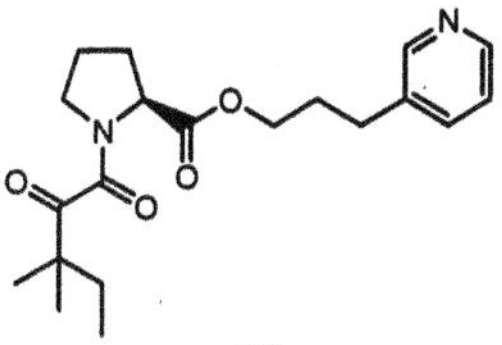

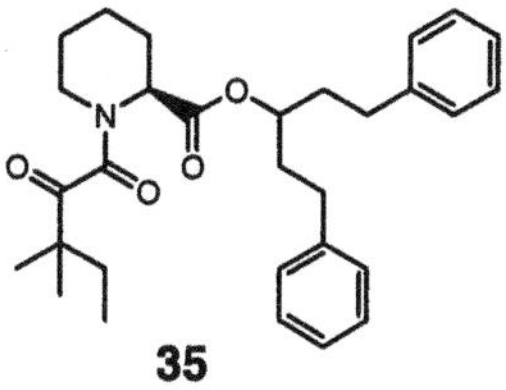

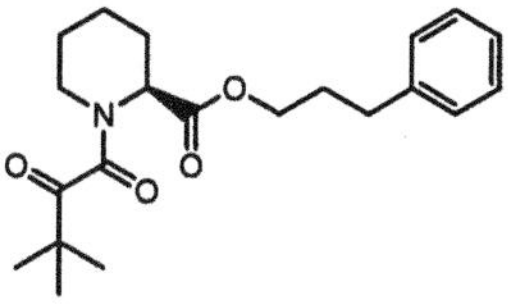

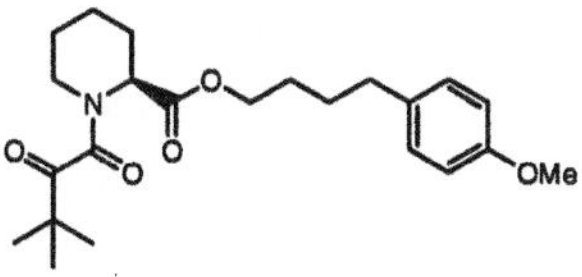

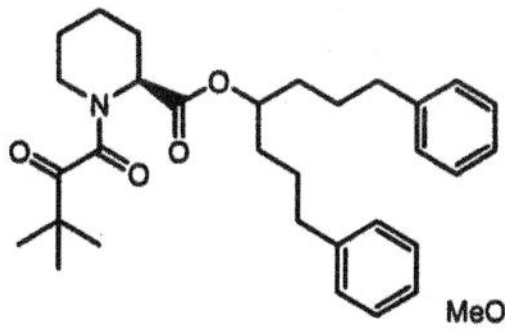

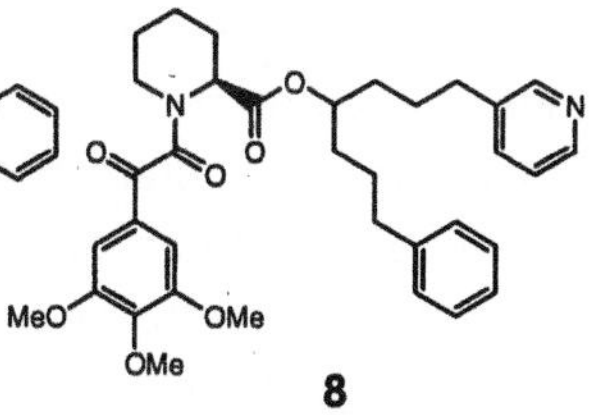

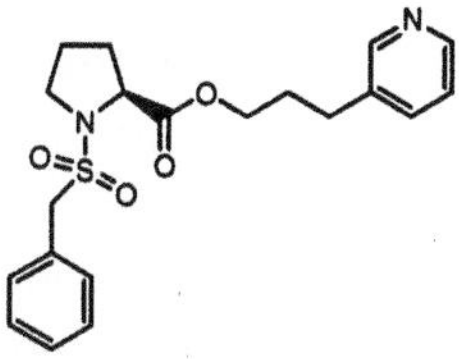

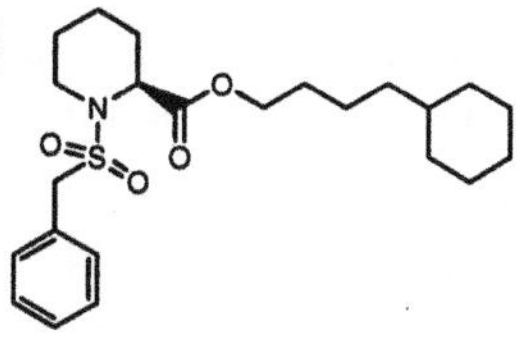

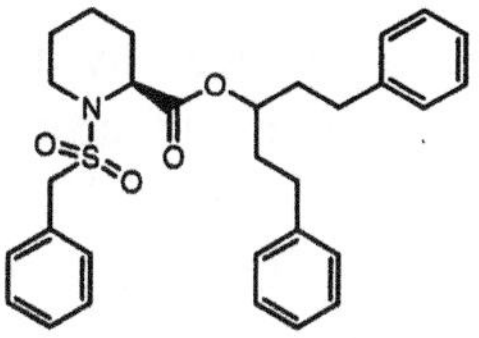

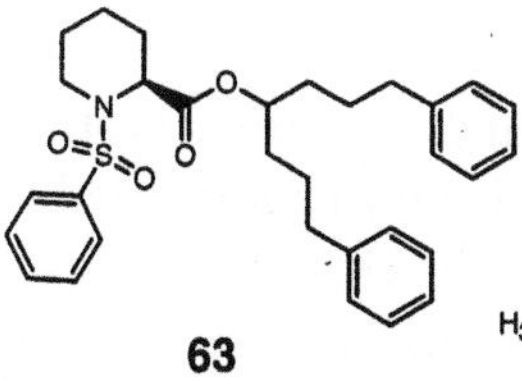

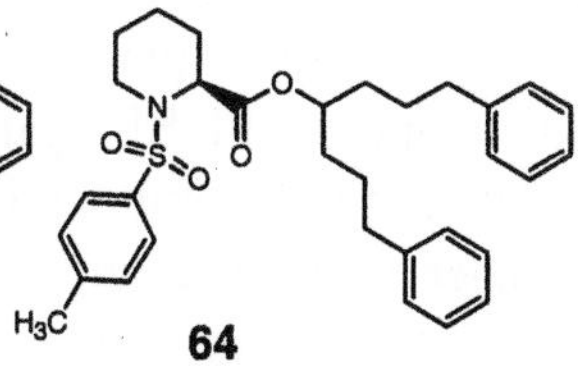

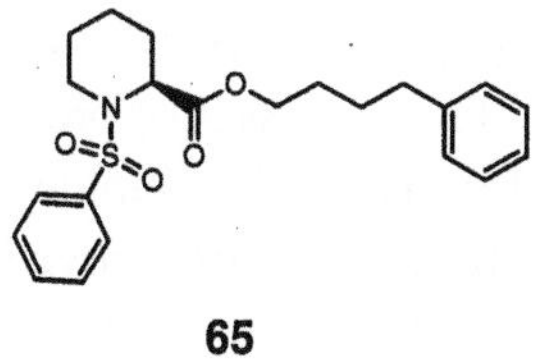

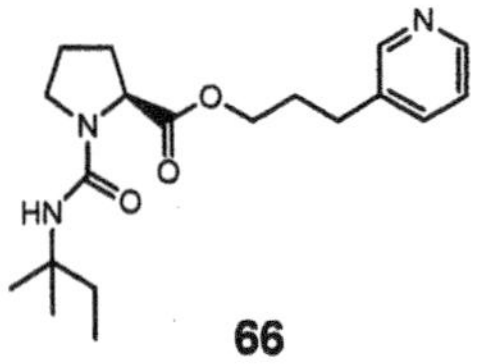

Figure 14. Continued.

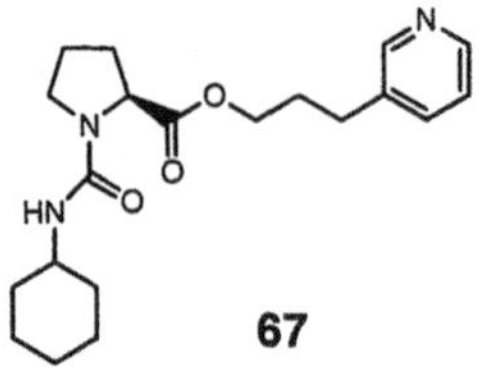

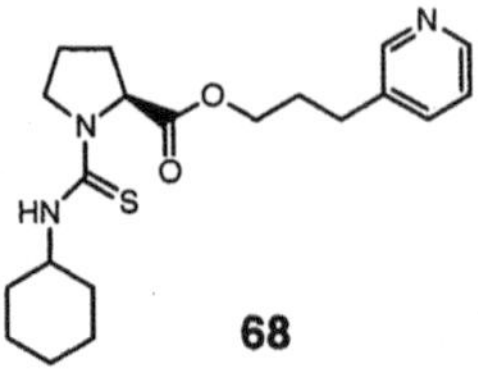

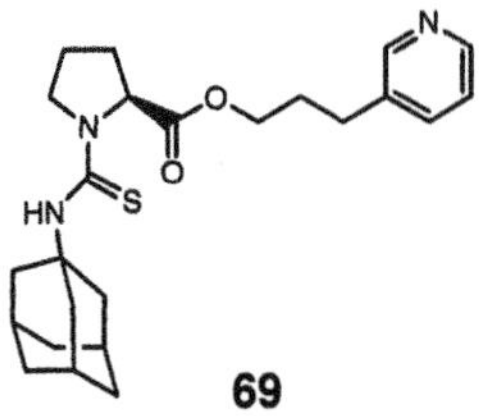

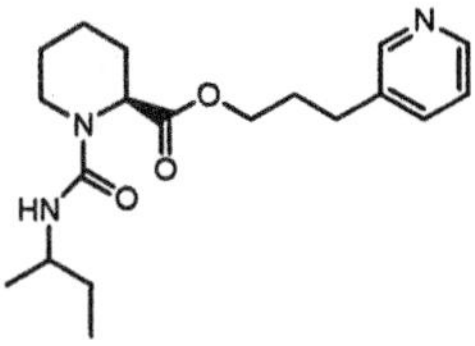

The extremely potent neurotrophic effects of many of these compounds make it clear that the nerve growth activity of FK506 does not depend upon an intact effector domain. Many of these compounds are remarkably potent, producing striking increases in process extension at picomolar concentrations. Neurite outgrowth in these assays was scored by counting the number of processes longer than the DRG explant diameter 48 hours after drug treatment. Neurites elicited by the actions of these compounds are indistinguishable from those produced by treatment with nerve growth factor: both more processes, and longer processes, are produced by increasing doses of the drugs, and the neurites form a dense, highly arborized network with a morphology comparable to that produced by NGF treatment. Figure 15 presents a series of photomicrographs of compound **10** (GPI 1046; Figure 14) causing process extension at increasing doses. In these cultures, the maximum number of processes per explant elicited by **10** averaged 140; this effect is comparable to the maximal effect produced by treatment with 50 ng/mL of NGF (ca. 140–150 processes/explant).

Other classes of previously reported FKBP ligands are also neurotrophic. The α-ketoamide moiety in these compounds may be replaced by sulfonyl or carbamoyl linkages with retention of in vitro and in vivo neurotrophic effects (Figure 14). Like the compounds in Table 4, these molecules were shown to be effective ligands for, and inhibitors of, FKBP12 (Tables 5 and 6). Representative compounds were also potent neurotrophic agents in vitro, promoting neurite outgrowth from cultured sensory neurons with nanomolar or subnanomolar ED_{50}s.[265]

Table 4. In Vitro Activities of N-Glyoxyl Prolyl and Pipecolyl Esters

Compound	n	R'	R	K_i, nM	ED_{50}, nM
9	1	1,1-Dimethylpropyl	3-Phenylpropyl	42	??
17	1	"	3-Cyclohexylpropyl	94	0.26
10	1	"	3-(3-Pyridyl)propyl	7.5	0.05
18	1	"	3-(2-Pyridyl)propyl	195	0.08
19	1	"	3-(2,5-Dimethoxyphenyl)propyl	250	2
20	1	"	3-(2,5-Dimethoxyphenyl)prop-2-enyl	450	0.8
22	1	"	(3,4,5)-Trimethoxyphenylethyl	120	0.015
23	1	"	3-(3,4-Methylenedioxyphenyl)propyl	170	10
24	1	Cyclohexyl	3-Phenylpropyl	82	0.13
25	1	Cyclohexyl	3-(3-Pyridyl)propyl	9	2
26	1	tert-Butyl	3-Phenylpropyl	95	0.025
27	1	tert-Butyl	3-(3-Pyridyl)propyl	3	0.014
28	2	OMe	Et	>10,000	>10,000
29	2	OMe	Benzyl	8000	>10,000
30	2	OMe	4-Cyclohexylbutyl	6000	>10,000
31	2	OMe	3-Cyclopentylpropyl	>10,000	>10,000

57

Table 4. Continued

Compound	n	R'	R	K_i nM	ED_{50} nM
32	2	OMe	3-Cyclohexylpropyl	>10,000	>10,000
33	2	Phenyl	Me	>10,000	>10,000
2	2	1,1-Dimethylpropyl	Et	660	5000
3	2	"	3-Phenylpropyl	110	300
4	2	"	(3,4,5)-Trimethoxyphenylpropyl	12	80
34	2	"	3-(3-Pyridyl)propyl	130	8.5
35	2	"	1,5-Diphenyl-3-pentyl	20	0.016
36	2	"	1,7-Diphenyl-4-heptyl	30	5
37	2	"	4-(4-Methoxyphenyl)-6-phenyl-3-hexyl	15	0.17
38	2	"	4,4-Diphenylbutyl	8.3	0.029
39	2	"	4-(4-Methoxyphenyl)-1-butyl	60	43
40	2	"	3-(2,5-Dimethoxyphenyl)propyl	12	2.5
41	2	"	3-Cyclohexylpropyl	170	1
5	2	"	(1R)-1,3-Diphenylpropyl	10	0.30
42	2	tert-Butyl	3-Phenylpropyl	2	0.085
43	2	"	4-(4-Methoxyphenyl)-1-butyl	24	0.002

58

No.	R		Substituent	Value 1	Value 2
44	"	2	1,7-Diphenyl-4-heptyl	5	0.085
45	"	2	3-Cyclohexylpropyl	32	0.29
46	"	2	2-Adamantyl	78	0.11
47	Cyclohexyl	2	3-Phenylpropyl	210	0.82
48	"	2	3-Cyclohexylpropyl	760	1
49	"	2	4-(4-Methoxyphenyl)-1-butyl	103	6
50	Phenyl	2	3-(3-Pyridyl)propyl	725	0.8
52	"	2	1,7-Diphenyl-4-heptyl	3	0.15
53	"	2	1-Phenyl-7-(2-pyridyl)-4-heptyl	50	0.063
54	(3,4,5)-Trimethoxy-phenyl	2	4-Cyclohexylbutyl	40	0.031
55	"	2	4-(4-Methoxyphenyl)-1-butyl	13	0.03
56	"	2	(3-Phenoxy)benzyl	18	0.18
57	"	2	3-(3-Indolyl)propyl	17	0.06
58	"	2	1,7-Diphenyl-4-heptyl	1	0.61
59	"	2	1-Phenyl-6-(3-pyridyl)-4-hexyl	3	0.95
60	"	2	1-Phenyl-7-(2-pyridyl)-4-heptyl	1	0.05
8	"	2	1-Phenyl-7-(3-pyridyl)-4-heptyl	0.5	n.d.

Table 5. In Vitro Activity of N-Sulfonyl Esters

Compound	n	R'	R	K_i, nM	ED_{50}, nM
61	1	Benzyl	3-(3-Pyridyl)propyl	72	nd
12	2	Benzyl	4-Phenylbutyl	34	0.031
62	2	Benzyl	3-(1,5-Diphenyl)pentyl	107	0.133
63	2	Phenyl	4-(1,7-Diphenyl)heptyl	332	1.0
64	2	4-Methyl-phenyl	4-Phenylbutyl	504	nd
65	2	Phenyl	4-Phenylbutyl	470	nd

Table 6. In Vivo Activity of N-Carbamoyl Pipecolyl and Prolyl Esters

Compound	n	X	R'	R	K_i, nM	ED_{50}, nM
66	1	O	1,1-Dimethylpropyl	3-(3-Pyridyl)propyl	742	1.0
67	1	O	Cyclohexyl	3-(3-Pyridyl)propyl	1482	nd
68	1	S	Cyclohexyl	3-(3-Pyridyl)propyl	131	0.30
69	1	S	2-Adamantyl	3-(3-Pyridyl)propyl	116	0.14
70	2	O	2-Methylbutyl	3-(3-Pyridyl)propyl	70	0.07

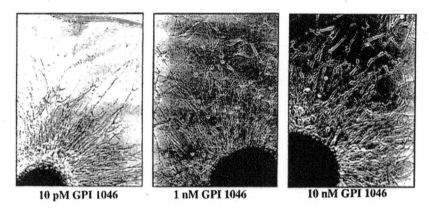

| 10 pM GPI 1046 | 1 nM GPI 1046 | 10 nM GPI 1046 |

Figure 15. FKBP ligand GPI 1046 promotes neurite outgrowth from cultured sensory neurons in a dose-dependent manner.

VI. THERAPEUTIC UTILITY OF IMMUNOPHILIN LIGANDS IN TREATING NEUROLOGICAL DISORDERS

Following the discovery that nonimmunosuppressive immunophilin ligands retained the neurotrophic properties of FK506, a number of compounds have been evaluated for efficacy in several animal models of neurodegenerative disorders. FK506 had already been shown to promote recovery of damaged peripheral nerves; several reports now detail the ability of compounds in Tables 4 and 5 to promote regeneration of both central and peripheral neurons.

Compounds **3**, **4**, **8**, and **10** were shown to promote regrowth of damaged peripheral nerves in vivo in rats. Treatment of rats with FKBP ligands for 18 days following lesioning of the sciatic nerve produced significant regeneration of damaged nerves, resulting in both more axons and axons with a greater degree of myelination in drug-treated animals relative to controls. At doses of 30 mg/kg/day, administered subcutaneously, compounds **3** and **4** were as effective as FK506 (Table 6).[263] Lesioning of the animals treated only with vehicle caused a significant decrease in axon number and degree of myelination compared to controls; treatment with **3** and **4** augmented both regrowth and remyelination of the damaged nerve. Compound **10** produced similar results at doses of 3 and 10 mg/kg/day, administered subcutaneously (Table 7).[266] Particularly striking were the effects observed on myelination levels. As quantitated by staining with anti-myelin basic Ig, myelin sheaths were not evident in lesioned, nondrug-treated animals, while animals receiv-

Table 7. FKBP Ligand-Induced Regeneration of Lesioned Sciatic
Nerves[a]

Treatment[b]	Myelination	Cross sectional Area (mm²)	Axonal Diameter (mm)
Sham	55.98	8.69 ± 0.50	2.62 ± 0.18
Lesion/vehicle	2.95	2.54 ± 0.38	1.40 ± 0.20
Lesion/Cpd. **3** (30 mg/kg)[c]	21.13	5.04 ± 0.40	2.06 ± 0.36
Lesion/Cpd. **4** (30 mg/kg)[c]	n.d.	3.84 ± 0.28	1.81 ± 0.12
Sham	63.44 ± 8.50	12.04 ± 0.95	3.46 ± 0.15
Lesion/vehicle	3.18 ± 1.59	4.61 ± 0.32	1.94 ± 0.07
Lesion/Cpd. **10** (3 mg/kg)[c]	21.66 ± 5.94	8.62 ± 0.45	2.66 ± 0.08
Lesion/Cpd. **10** (10 mg/kg)[c]	24.31 ± 7.61	8.87 ± 0.34	2.74 ± 0.04

Notes: [a]Data for compounds **3** and **4** taken from ref. 263; data for compound **10** taken from ref. 266.

[b]Rats with lesioned sciatic nerves were treated with the indicated compounds, and sacrificed on the 18th day of the experiment. Axonal diameter and cross-sectional area were quantitated by anti-neurofilament staining, and myelin levels were quantitated by myelin basic protein–immunoreactive stain density.

[c]Compounds were administered subcutaneously in Intralipid, at the indicated doses, once a day for 18 days.

ing **10** showed substantial myelin layering in the injured nerves, and had myelin levels 7–8 times higher than lesion/vehicle animals.

Also significant was the demonstration that the compounds did not appear to affect healthy nerves. Experiments with protein growth factors have shown that they may induce abnormal neuronal sprouting.[267] Treatment with 40 mg/kg/day, sc, of **10** for 18 days did not cause abnormal axonal growth or altered myelination patterns in the sciatic nerves of healthy, nonlesioned rats.

Similar results in the rat sciatic nerve model were reported by Gold et al. for compound **8** (V-10,367; Figure 14).[268] Evaluation of functional and morphological recovery was done as described previously for FK506. Daily sc injections of V-10,367 (400 mg/kg) accelerated the onset of functional recovery relative to controls, and this improvement was reflected morphologically in the observation of larger regenerating axons in the drug-treated animals. V-10,367 was effective following oral administration at doses as low as 5 mg/kg/day.[269] Consistent with previous studies, animals receiving V-10,367 orally had larger axonal calibers and more myelinated fibers distal to the site of the lesion. Functional recovery in the drug-treated animals was reduced to 13 days from 17 days in control animals, an effect superior to that produced by sc administration of FK506.

More recent reports describe the ability of immunophilin ligands to promote recovery following peripheral neuropathy in rodents induced by treatment with streptozotocin or cisplatin. In the streptozotocin-induced diabetic rat, animals treated with compound **10** (GPI 1046) for 6 weeks (10 mg/kg/day sc) showed behavioral improvement, evaluated using the hot plate and tail flick apparatus to determine changes in pain sensation.[270] Drug-treated animals manifested improved latency in the tail flick test, suggesting that treatment with **10** ameliorated the behavioral sequelae of diabetic sensory neuropathy. Hypoalgesia (decreased sensation to pain) in cisplatin-treated mice was also evaluated using the tail flick assay. Hypoalgesia was apparent after 4 weeks of cisplatin treatment and became progressively worse with continued cisplatin treatment.[271] Concurrent treatment with GPI 1046 (**10**; 20 mg/kg/day sc) either blocked the development of hypoalgesia or promoted a restoration of normal sensitivity to pain.

As discussed in an earlier section, the nigrostriatal pathway in the brain was found to be particularly enriched in FKBP12. This pathway is responsible for controlling motor movements. The axons of the neurons comprising this pathway rise from the cell bodies in the substantia nigra pars compacta, through the medial forebrain bundle, and terminate in the striatum where the terminals release dopamine as a neurotransmitter. Parkinson's disease is a serious neurodegenerative disorder resulting from degeneration of this motor pathway and subsequent decrease in dopaminergic transmission.[272] *N*-Methyl-4-phenyl-1,2,3,6-tetrahydropyridine (MPTP) is a neurotoxin which selectively destroys dopaminergic neurons;[273] lesioning of the nigrostriatal pathway in animals with MPTP has been utilized extensively as an animal model of Parkinson's disease. FKBP ligands have been shown to possess potent neuroprotective and neuroregenerative properties in this disease model following either systemic or oral administration.

In a study utilizing a protective or "concurrent" protocol, mice were treated with MPTP and FKBP12 ligands, administered subcutaneously, concurrently for 5 days.[264] Test compounds were given for an additional 5 days, and after 18 days the animals were perfused and the brains were fixed, cryoprotected, and sectioned. Staining with an antibody against tyrosine hydroxylase (TH) was used to quantitate survival of dopaminergic neurons. In animals treated with MPTP and vehicle, a substantial loss of 60–70% of functional dopaminergic terminals was observed as compared to nonlesioned animals. Lesioned animals receiving compounds **35**, **42**, **43**, and **44** (Figure 14) (40 mg/kg, sc) concurrently with

Table 8. FKBP12-Ligand-Induced Recovery of Striatal
Tyrosine Hydroxylase Immunostaining in MPTP-Treated
Mice

Compound	Dosage	% Recovery Striatal Innervation
10	40 mg/kg sc	67.50 ± 4.30
12	4 mg/kg sc	38.02 ± 4.00
35	40 mg/kg sc	71.60 ± 4.90
42	"	32.40 ± 3.98
43	"	58.90 ± 4.93
44	"	43.80 ± 8.65
61	4 mg/kg sc	44.31 ± 4.73
64	"	44.16 ± 3.37
65	"	29.22 ± 3.00
67	"	59.79 ± 4.57
68	"	56.13 ± 4.74
69	"	52.32 ± 3.89
70	"	27.47 ± 2.65

MPTP showed a striking recovery of TH-stained striatal dopaminergic terminals, as compared with controls, demonstrating the ability of the compounds to block the degeneration of dopaminergic neurons produced by MPTP (Table 8). These compounds are simple mimetics of the FKBP-binding portion of FK506, and their ability to promote recovery of dopaminergic innervation in this model was a critical demonstration that the neurotrophic effects of FK506 resided within this structural array of the immunosuppressant.

Other FKBP ligands from Tables 4 and 5 were also shown to be effective in this model, including prolyl ester **10**,[266] sulfonamides **12, 61, 64,** and **65**,[222,265] and ureas **67–70**.[222,265] The data for these compounds in this "concurrent dosing" paradigm is gathered in Table 7.

The dose-response curve of compound **10** following systemic administration was similar to that reported by Gold for FK506 in the rat sciatic nerve model. A maximum efficacy of 86% recovery of tyrosine hydroxylase positive striatal fibers was found at 20 mg/kg sc (Figure 16).[266] Again, the drug did not appear to have an adverse effect on healthy neurons; animals receiving 40 mg/kg/day subcutaneously of **10** for 10 days did not show evidence of intranigral sprouting or increased striatal dopaminergic innervation.

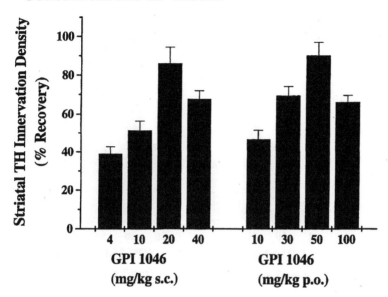

Concurrent MPTP Model

Figure 16. Dose-response curves for recovery of striatal dopaminergic innervation in mice by GPI 1046 (compound 10, Table 4) in the concurrent dosing paradigm. Data for subcutaneous administration are taken from ref. 266 and data for oral administration from refs. 274 and 275.

The powerful therapeutic potential of these compounds was further extended by the report that GPI 1046 (**10**) was effective in the mouse MPTP model following oral administration.[274,275] When administered concurrently with MPTP at a dose of 50 mg/kg po, the drug produced a remarkable recovery of striatal dopaminergic innervation to greater than 90% of control levels (Figure 16). Furthermore, GPI 1046 and related compounds were effective in restoring striatal innervation when administered subsequent to lesioning with MPT.[266,274,275] In these experiments, mice received treatment with FKBP ligands beginning 3 days following cessation of MPTP treatment, at which time the reduction in striatal TH-positive fiber density is maximal. As shown in Figure 17, post-MPTP systemic (sc) administration of **10** produced a dose-dependent increase in striatal innervation as visualized by tyrosine hydroxylase immunohistochemistry (Figure 18). Administered orally, GPI 1046 also regenerated striatal dopaminergic terminals post-MPTP lesioning. At an oral dose of

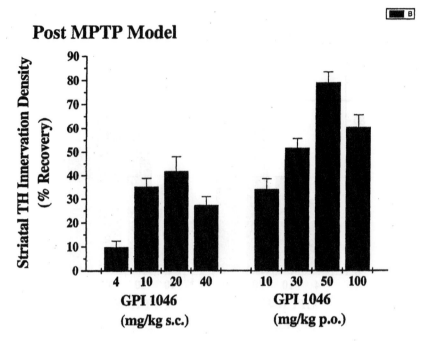

Figure 17. Dose-response curves for recovery of striatal dopaminergic innervation in mice by GPI 1046 (compound 10, Table 4) in the post-MPTP dosing paradigm. Data for subcutaneous administration are taken from ref. 266 and data for oral administration from refs. 274 and 275.

50 mg/kg, the compound regenerated striatal terminals to about 80% of control levels (Figure 17), suggesting a remarkable ability of the compounds to regenerate damaged central neuronal pathways. It was also shown that the dopaminergic terminals in these drug-treated animals were functional. Treatment of the animals with MPTP caused an approximately 50% depletion of striatal dopamine; treatment with GPI 1046, either systemically or orally, produced a significant restoration of dopamine and its metabolites in the striatum.[266,274] The experimental results suggest that the compounds stimulate axonal sprouting from spared projections, regenerating functional terminals. These compounds appear to be the first orally active small molecules capable of doing so, and may therefore represent a powerful new approach to the treatment of neurodegenerative disorders.

Intranigral injection of 6-OHDA (a severe neurotoxin which, like MPTP, destroys neurons which release dopamine as a neurotransmit-

TH Immunostaining in Striatum of Post-MPTP Treated Mice

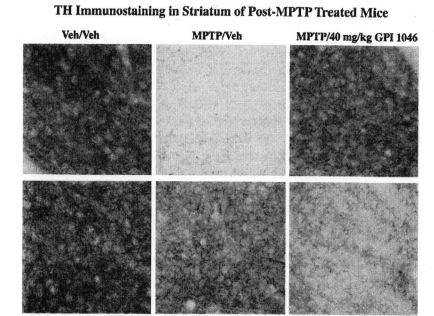

Veh/Veh   **MPTP/Veh**   **MPTP/40 mg/kg GPI 1046**

MPTP/20 mg/kg GPI 1046   **MPTP/10 mg/kg GPI 1046**   **MPTP/4 mg/kg GPI 1046**

Figure 18. Tyrosine hydroxylase immunostaining in representative brain sections of mice in the post-MPTP protocol.

ter[276]) into one side of the brains of rats causes a substantial destruction of dopaminergic neurons in the side of the brain receiving the injection (the ipsilateral side), while leaving the other (contralateral) side unaffected. Treatment of such animals with amphetamine produces a circling or rotating behavior in a direction towards the lesioned side.[277] In animals lesioned with 6-OHDA, a 90–95% reduction in TH-positive striatal fiber density was observed. Subcutaneous administration of GPI 1046, 10 mg/kg/day for 5 days beginning 1 week following the lesion, restored striatal dopaminergic terminals to greater than 40% of control levels.[266] Remarkably, significant reinnervation was observed even when drug treatment was delayed for up to 4 months following production of the lesion.[278] As was the case in the MPTP mouse model, recovery of striatal TH immunostaining was accompanied by restoration of striatal levels of dopamine and dopamine metabolites. Morphological analysis of the brains of the animals indicated that the reinnervation of the striatum was due to intrastriatal sprouting of spared fibers from the substantia nigra compacta and VTA. The morphological recovery was reflected in a reduction of the motor abnormalities of the drug-treated animals. Amphetamine-induced rotational behavior was significantly attenuated in animals receiving GPI 1046. Taken together, the data suggest that FKBP ligands promote biochemical, morphological, and behavioral recovery in the animals in these experiments.

In a separate report, sprouting of residual cholinergic axons following partial transection of the fimbria fornix in rats was induced by treatment with GPI 1046.[279] A 75–80% deafferentation of the hippocampus is caused by the transection. Subcutaneous administration of GPI 1046, 10 mg/kg/day for 14 days, caused substantial recovery of cholinergic innervation in the hippocampus, as visualized by choline acetyl transferase (ChAT) immunostaining. The increased level of innervation was observed to result from sprouting of spared residual processes in the CA1, CA3, and dentate gyrus regions of the hippocampus, producing 62% (CA1), 48% (CA3), and 30% (dentate) recovery of ChAT+ fiber density in these regions.

Cholinergic deafferentation has been suggested to play a role in the cognitive deficits associated with Alzheimer's disease.[280] The results described here suggest that FKBP ligands may be of utility in treating learning and memory disorders. A recent abstract describes the effects of GPI 1046 on learning and memory in aged mice.[281,282] The performance of young and aged mice in the Morris watermaze was evaluated, and aged rodents were sorted into two groups on the basis of their performance relative to young mice: "aged impaired" or "aged nonimpaired." Daily treatment of aged rodents with vehicle or GPI 1046 (10 mg/kg/day) was initiated 3 days after acquisition training ended and was continued through retention testing, which was done after 3 weeks of dosing. Animals in the aged-impaired vehicle-treated group performed significantly more poorly than aged-impaired GPI 1046-treated animals and "young" animals. There was no statistical difference in the retention phase between young rodents and aged-impaired rodents that had received the immunophilin ligand, indicating that systemic treatment with the compound enhanced spatial memory performance of mice with age-related spatial memory impairments.

In addition to central dopaminergic and cholinergic neurons, serotonergic neurons are also affected by FKBP ligands. Rats treated with the neurotoxin *para*-chloroamphetamine (PCA) suffer destruction of neurons which use serotonin as a neurotransmitter. Rats treated with GPI 1046, administered at the time of PCA lesioning and for 14 days thereafter, were compared with rats receiving PCA or vehicle alone. PCA lesioning caused a reduction of cortical density of serotonin fibers to 10% of control levels, while animals receiving GPI 1046 (1–40 mg/kg/day, sc) had significantly greater serotonin fiber innervation in the somatosensory cortex than lesioned, nondrug-treated rodents.[266,283]

GPI 1046 was also reported to protect organotypic spinal cord neurons from exitotoxicity due to loss of glutamate transport, providing a powerful neuroprotective effect.[284] The mechanism of neuroprotection appeared to be mediated through FKBP12, and induction of upregulation of two glutamate transporter subtypes (GLAST and GLT-1) was observed.

The neuroprotective effects of FKBP ligands following transection of the optic nerve in rats has been studied.[285] Retinal ganglion cells were labeled with fluorogold (FG), and the optic nerves were transected 5 mm behind the globe 4 days later. Groups of animals were treated with GPI 1046 (10 mg/kg/day sc) or vehicle for 28 days. On the 90th day of the experiment all treated animals and controls were sacrificed. Treatment with the FKBP ligand provided moderate protection of retinal ganglion cells after transection, but produced striking decreases in myelin and axon degeneration in the proximal stump (portion of the severed nerve still attached to the eye) and prevented myelin degeneration in the distal stump (portion of the optic nerve disconnected from the eye). These results indicate that treatment with the neuroimmunophilin ligand produced a fundamental alteration in the pathological process following injury to the optic nerve.

Taken together, the results surveyed here suggest a remarkable ability of FKBP ligands to block degeneration and promote regeneration of damaged nerves. The ability of these compounds to promote functional and structural in a variety of central and peripheral nerve types suggests therapeutic utility in a wide range of nervous system disorders, including Parkinson's and Alzheimer's diseases, peripheral neuropathies, spinal cord and traumatic head injury, and brain damage from stroke. Given the intense interest in this area by several companies, it is likely that one or more neuroimmunophilin ligands will enter human testing within the next year or two.

VII. SUMMARY AND CONCLUDING REMARKS

In the 15 years since the discovery of cyclophilin there has been an enormous proliferation of known peptidylprolyl isomerases. The members of the FKBP and cyclophilin families, known collectively as the immunophilins, are widely distributed throughout mammals, and have been observed to interact with numerous proteins and signaling pathways. Far from functioning as simply foldases, the immunophilins are involved in protein trafficking and modulation of signal transduction pathways. Several examples of roles played by immunophilins in assembling and modulating the activity of multicomponent macromolecular complexes have been described, with others no doubt to follow. It seems clear that we are just beginning to appreciate the broad role played by these fascinating proteins in cellular processes.

It is interesting to note that the enzymatic activity of the immunophilins at times appears to be irrelevant in the context of their scaffolding functions. As seen in the interaction of FKBP12 with ryanodine and IP_3 receptors, rotamase activity per se is not important, though the interactions with the target proteins do occur through the rotamase domain. In the context of these interactions, the rotamase activity may essentially be an artifact. We suggest that some immunophilins may function as adapter proteins, with the FKBP and/or cyclophilin domains functioning as molecular sockets for the assembly of protein complexes, in the manner of a tinkertoy set. What may be important in this context is not the enzymatic activity itself, but the geometrical recognition element conferred by the substrate specificity of the enzymatic active site/binding domain. As discussed in a previous section, computational studies and analysis of protein substrates bound to FKBP12 and CyPA suggest that the two immunophilins bind different types of β-turns in proteins. The immunophilins are thus particularly suited to bind to various proteins, such as membrane-bound receptors, that contain cytosolic reverse turns. The FKBP and cyclophilin domains discussed in the structural biology section may be analogous to the SH2 and SH3 domains found in adapter molecules such as Grb2.[286] The core rotamase domains provide a means of recognizing specific types of reverse turns, while changes in the electrostatic and steric environment surrounding the domain may serve to confer specificity for protein interactions. The recent observation of proteins containing both FKBP and cyclophilin domains suggests that these domains may be utilized in a modular approach to construct customized docking molecules for macromolecular complex assembly. Similar domains may be found in other proteins not presently thought of as immunophilins; for example, the scaffolding protein AKAP78, which binds protein kinase A and calcineurin and targets them to specific subcellular sites, contains a calcineurin-binding site which resembles FKBP12.[287]

Surely one of the most exciting discoveries in the immunophilin field is the remarkable finding that small molecule, nonimmunosuppressive immunophilin ligands promote regeneration of damaged nerves. The discovery of agents which would effect repair of degenerating nervous tissue has long been a Holy Grail of neuroscience. The studies reviewed here suggest that immunophilin ligands may possess a broad spectrum of activity in the peripheral and central nervous systems, raising the tantalizing possibility of therapeutic utility in a variety of neurodegenerative disorders. Although neurotrophic proteins are being explored by several biotechnology companies for therapeutic utility,[272] peptidic growth factors are severely limited in their utility because of their lack of oral bioavailability and inability to penetrate the central nervous system, and have been observed to cause abnormal neuronal sprouting.[288] In contrast, immunophilin ligands such as those described here are active upon oral administration and readily penetrate the blood–brain barrier. Even more striking is their lack of effect on normal, healthy neurons. This latter property may be related to the observation that FKBP12 is rapidly upregulated in damaged nerves and transported to the site of injury.

The field of neuroimmunophilins is still in its infancy, and many basic issues are as yet unresolved. The conflicting reports regarding the neurotrophic actions of cyclosporin A are one example. Gold has reported that cyclosporin A does not promote neurite outgrowth in SH-SY5Y cells even at concentrations of 1 μM;[259] however, Steiner et al. described potent neurotrophic effects for CsA in both PC12 cells and chick dorsal root ganglion cells.[263] Gold's group also reported that, in a comparative study with FK506, CsA did not increase the regeneration rate of axons in rats with lesioned sciatic nerves,[289] and Yagita et al. described a study in which CsA was found to be ineffective in protecting hippocampal and cerebral neurons against transient global ischemia.[252] On the other hand, a recent report states that systemic administration of CsA to rats with compromised blood–brain barriers resulted in increased nigral dopaminergic neuronal outgrowth.[290] Future studies will hopefully clarify the nervous system effects of cyclophilin ligands.

Many unanswered questions remain regarding the nervous system actions of neuroimmunophilin ligands. The mechanism of action of the neural effects of FKBP ligands remains to be elucidated, and will undoubtedly be the focus of much research in the near term. The lack of correlation between inhibition of FKBP12 rotamase activity and in vitro neurotrophic potency was noted in a previous section. This observation, together with the fact that many of the compounds exert their effects at concentrations at which only a small part of the FKBP12 present in a typical neuron would be inhibited, indicates that FKBP12 inhibition is not the source of neurotrophic activity. It is possible that other FKBPs, present in lower concentrations in nerve cells, may be involved. Injection of FK506 subcutaneously into monkeys caused increased expression of hsp70, a heat shock protein, in neurons in animals.[291] Regions of the nervous system so affected included the spinal cord, dorsal root ganglion, and several regions of the brain. Noting these results, Gold has suggested FKBP52 as a potential target of the compounds.[259] Administration of FK506 has also been found to increase messenger RNA levels for GAP-43 in neurons. Other proteins containing FKBP or FKBP-like domains may possibly be responsible for mediating some or all of the neuroprotective and neuroregenerative actions described here.

The recurring role of immunophilins as components of macromolecular complexes has been discussed, and it is possible that binding of the compounds described herein to an FKBP domain promotes formation of an activated complex leading to a gain of function. The ability of small molecule ligands to induce changes in the local conformation of FKBP12 has been discussed in an earlier section, and the existence of an extended binding domain which may be alternately exposed or concealed by the hingelike movement of the 80s loop has been postulated. Binding of compounds such as those in Figure 14 may enhance protein–protein interactions; alternatively, the ligands may bind to a preexisting multicomponent complex and alter its function.

The application of modern tools of molecular biology and biochemistry, such as affinity chromatography and labeling, the yeast two-hybrid system, and use of antisense techniques, to these problems will undoubtedly shed light on these issues and identify potential protein targets in the relevant neurotrophic pathways. In addition to expanding the basis for therapeutic drug design, these studies will illuminate areas of fundamental neurobiology which are currently shrouded in mystery. Once again, the immunophilins and their ligands seem poised to usher in exciting new areas of research at the interface of chemistry and biology.

REFERENCES

1. For a more comprehensive picture of this topic, the reader is referred to the following reviews: (a) Schreiber, S. L. *Science* **1991**, *251*, 283–287. (b) *Angew.*

Chem., Intl. Ed. Engl. **1992**, *31*, 384–400. (c) Armistead, D.M.; Harding, M.W. *Ann. Rep. Med. Chem.* **1993**, *28*, 207–215.

2. Fischer, G.; Bang, H.; Mech, C. *Biomed. Biochim. Acta* **1984**, *43*, 1101–1112.

3. Fischer, G.; Bang, H. *Biochim. Biophys. Acta* **1984**, *828*, 39–42.

4. Borel, J. F. *Pharmacol. Rev.* **1989**, *41*, 259–371.

5. Handschumacher, R. E.; Harding, M. W.; Rice, J.; Drugge, R. J.; Speicher, D. W. *Science* **1984**, *226*, 544–546.

6. Fischer, G.; Wittmann-Liebold, B.; Lang, K.; Kiefhaber, T.; Schmid, F. X. *Nature* **1989**, *337*, 476–478.

7. Takahashi, N.; Hayano, T.; Suzuki, M. *Nature* **1989**, *337*, 473–475.

8. (a) Sigal, N. H.; Dumont, F. J. *Ann. Rev. Immunol.* **1992**, *10*, 519–560. (b) Kunz, J.; Hall, M. N. *Trends Biochem. Sci.* **1993**, *18*, 334–338.

9. Harding, M. W.; Galat, A.; Uehling, D. E.; Schreiber, S. L. *Science* **1989**, *341*, 761–763.

10. Siekerka, J. J.; Hung, S. H. Y.; Poe, M.; Lin, S. C.; Sigal, N. H. *Nature* **1989**, *341*, 755–757.

11. Tocci, M. J.; Matkovich, D. A.; Collier, K. A.; Kwok, P.; Dumont, F.; Lin, C. S.; DeGudicibus, S.; Siekerka, J. J.; Chin, J.; Hutchinson, N. I. *J. Immunol.* **1989**, *143*, 718–726.

12. Lin, C. S.; Boltz, R. C.; Siekerka, J. J.; Sigal, N. H. *Cell. Immunol.* **1991**, *133*, 269–284.

13. Bierer, B. E.; Mattila, P. S.; Standaert, R. F.; Herzenberg, L. A.; Burakoff, S. J.; Crabtree, G.; Schreiber, S. L. *Proc. Natl. Acad. Sci. USA* **1990**, *87*, 9231–9235.

14. Dumont, F. J.; Staruch, M. J.; Koprak, S. L.; Melino, M. R.; Sigal, N. H. *J. Immunol.* **1990**, *144*, 251–258.

15. Dumont, F. J.; Melino, M. R.; Staruch, M. J.; Koprak, S. L.; Fisher, P. A.; Sigal, N. H. *J. Immunol.* **1990**, *144*, 1418–1424.

16. Bierer, B. E.; Somers, P. K.; Wandless, T. J.; Burakoff, S. J.; Schreiber, S. L. *Science* **1990**, *250*, 556–559.

17. Somers, P. K.; Wandless, T. J.; Schreiber, S. L. *J. Am. Chem. Soc.* **1991**, *113*, 8045–8056.

18. Dumont, F. J.; Staruch, M. J.; Koprak, S. L.; Siekerka, J. J.; Lin, C. S.; Harrison, R.; Sewell, T.; Kindt, V.; Beattie, T. R.; Wyvratt, M.; Sigal, N. H. *J. Exp. Med.* **1992**, *176*, 751–760.

19. Ocain, T. D.; Longhi, D.; Steffan, R. J.; Caccese, R. G.; Sehgal, S. N. *Biochem. Biophys. Res. Commun.* **1993**, *192*, 1340–1346.

20. Sigal, N. H.; Dumont, F.; Durette, P.; Siekierka, J. J.; Peterson, L.; Rich, D. H.; Dunlap, B. E.; Staruch, M. J.; Melino, M. R.; Koprak, S. L.; Williams, D.; Witzel, B.; Pisano, J. M. *J. Exp. Med.* **1991**, *173*, 619–628.

21. Liu, J.; Farmer, J. D. Jr.; Lane, W. S.; Friedman, J.; Weissman, I.; Schreiber, S. L. *Cell* **1991**, *66*, 807–815.

22. Friedman, J.; Weissman, I. *Cell* **1991**, *66*, 799–806.

23. Klee, C. B.; Crouch, T. H.; Krinks, M. H. *Proc. Natl. Acad. Sci. USA* **1979**, *76*, 6270–6273.

24. Flanagan, W. M.; Corthesy, B.; Bram, R. J.; Crabtree, G. R. *Nature* **1991**, *352*, 754–755.

25. Clipston, N. A.; Crabtree, G. R. *Nature* **1992**, *357*, 695–697.

26. O'Keefe, S. J.; Tamura, J.; Kincaid, R. F.; Tocci, M. J.; O'Neill, M. A. *Nature* **1992**, *357*, 692–694.

27. Kuo, C. J.; Chung, J.; Florentino, D. F.; Flanagan, W. M.; Blenis, J.; Crabtree, G. R. *Nature* **1992**, *358*, 70–73.

28. Chung, J.; Kuo, C. J.; Crabtree, G. R.; Blenis, J. *Cell* **1992**, *69*, 1227–1236.

29. Price, D. J.; Grove, J. R.; Calco, V.; Avruch, J.; Bierer, B. E. *Science* **1992**, *257*, 973–977.

30. Jayaraman, T.; Marks, A. R. *J. Biol. Chem.* **1993**, *268*, 25385–25388.

31. Morice, W. G.; Wiederrecht, G.; Brunn, G. J.; Siekerka, J. J.; Abraham, R. T. *J. Biol. Chem.* **1993**, *268*, 22737–22745.

32. Kunz, J.; Henriquez, R.; Schneider, U.; Deuter-Reinhard, M.; Movva, N. R.; Hall, M. N. *Cell* **1993**, *73*, 585–596.

33. Brown, E. J.; Albers, M. W.; Shin, T. B.; Ichikawa, K.; Keith, C. T.; Lane, W. S.; Schreiber, S. L. *Nature* **1994**, *369*, 756–758.

34. Sabatini, D. M.; Erdjument-Bromage, H.; Lui, M.; Tempst, P.; Snyder, S. H. *Cell* **1994**, *78*, 35–43.

35. Brown, E. J.; Beal, P. A.; Keith, C. T.; Chen, J.; Shin, T. B.; Schreiber, S. L. *Nature* **1995**, *377*, 441–446.

36. Chou, M. M.; Blenis, J. *Curr. Opin. Cell Biol.* **1995**, *7*, 806–814.

37. Brown, E. J.; Schreiber, S. L. *Cell* **1996**, *86*, 517–520.

38. Beretta, L.; Gingras, A.-C.; Svitkin, Y. V.; Hall, M.; Sonenburg, N. *EMBO J.* **1996**, *15*, 658–664.

39. Galat, A. *Eur. J. Biochem.* **1993**, *216*, 689–707.

40. Kay, J. E. *Biochem. J.* **1996**, *314*, 361–385.

41. Rahfield, J.-U.; Rucknagel, K. P.; Schelbert, B.; Ludwig, B.; Hacker, J.; Mann, K.; Fischer, G. *FEBS Lett.* **1994**, *352*, 180–184.

42. Rudd, K. E.; Sofia, H. J.; Koonin, E. V.; Plunkett, G.; Lazar, S.; Rouviere, P. R. *Trends Biochem. Sci.* **1995**, *20*, 12–14.

43. Standaert, R. F.; Galat, A.; Verdine, G. L.; Schreiber, S. L. *Nature* **1990**, *346*, 671–674.

44. Albers, M. W.; Walsh, C. T.; Schreiber, S. L. *J. Org. Chem.* **1990**, *55*, 4984–4986.

45. Sewell, T.; Lam, E.; Martin, M.; Leszyk, J.; Weidner, J.; Calaycay, J.; Griffith, P.; Williams, H.; Hung, S.; Cryan, J.; Sigal, N.; Wiederrecht, G. *J. Biol. Chem.* **1994**, *269*, 21094–21102.

46. Lam, E.; Martin, M. M.; Timerman, A. P.; Sabers, C.; Fleischer, S.; Lukas, T.; Abraham, R. T.; O'Keefe, S. J.; O'Neill, E. A.; Weiderrecht, G. *J. Biol. Chem.* **1995**, *270*, 26511–26522.

47. Jin, Y.-J.; Albers, M. W.; Lane, W. S.; Bierer, B. E.; Schreiber, S. L.; Burakoff, S. J. *Proc. Natl. Acad. Sci. USA* **1991**, *88*, 6677–6681.

48. Jin, Y.-J.; Burakoff, S. J.; Bierer, B. E. *J. Biol. Chem.* **1992**, *267*, 10942–10945.

49. Galat, A.; Lane, W. S.; Standaert, R. F.; Schreiber, S. L. *Biochemistry* **1992**, *31*, 2427–2434.

50. Murthy, J. N.; Goodyear, N.; Soldin, S. J. *Clin. Biochem.* **1997**, *30*, 129–133.

51. Peattie, D. A.; Harding, M. W.; Fleming, M. A.; DeCenzo, M. T.; Lippke, J. A.; Livingston, D. J.; Benasutti, M. *Proc. Natl. Acad. Sci. USA* **1992**, *89*, 10974–10978.

52. Lebeau, M.-C.; Massol, N.; Herrick, J.; Faber, L. E.; Renoir, J.-M.; Radanyi, C.; Baulieu, E.-E. *J. Biol. Chem.* **1992**, *267*, 4281–4284.

53. Yem, A. W.; Tommasselli, A. G.; Heinrikson, R. L.; Zurcher-Neely, H.; Ruff, V. A.; Johnson, D. A.; Deibel, M. R. *J. Biol. Chem.* **1992**, *267*, 2868–2871.

54. Harding, M.W.; Handschumacher, R. E.; Speicher, D. W. *J. Biol. Chem.* **1986**, *261*, 8547–8555.

55. Price, E. R.; Zydowsky, L. D.; Jin, M.; Baker, C. H.; McKeonand, F. D.; Walsh, C. T. *Proc. Natl. Acad. Sci. USA* **1991**, *88*, 1903–1907.

56. Schneider, H.; Charara, N.; Schmitz, R.; Wehrli, S.; Mikol, V.; Zurini, M. G.; Quesniaux, V. F.; Movva, N.R. *Biochemistry* **1994**, *33*, 8218–8224.

57. Connern, C. P.; Halestrap, A. P. *Biochem. J.* **1992**, *284*, 381–385.

58. Kieffer, L. J.; Thalhammer, T.; Handschumacher, R. E. *J. Biol. Chem.* **1992**, *267*, 5503–5507.

59. Anderson, S. K.; Gallinger, S.; Roder, J.; Frey, J.; Young, H. A.; Ortaldo, J. R. *Proc. Natl. Acad. Sci. USA* **1993**, *90*, 542–546.

60. Kofron, J. L.; Kuzmic, P.; Kishore, V.; Bonilla-Colon, E.; Rich, D. H. *Biochemistry* **1991**, *30*, 6127–6134.

61. Kofron, J. L.; Kuzmic, P.; Kishore, P.; Gemmecker, G.; Fesik, S. W.; Rich, D. H. *J. Am. Chem. Soc.* **1992**, *114*, 2670–2675.

62. Lang, K.; Schmid, F. X.; Fischer, G. *Nature* **1987**, *329*, 268–270.

63. Mucke, M.; Schmid, F.X. *Biochemistry* **1992**, *31*, 7848–7854.

64. Schonbrunner, E. R.; Mayer, S.; Tropschug, M.; Fischer, G.; Takahashi, N.; Schmid, F. X. *J. Biol. Chem.* **1991**, *266*, 3630–3635.

65. Hsu, V. L.; Handschumacher, R. E.; Armitage, I.A. *J. Am. Chem. Soc.* **1990**, *112*, 6745–6747.

66. Justice, R. M.; Kline, A. D.; Sluka, J. P.; Roeder, W. D.; Rodgers, G.H.; Roehm, N.; Mynderse, J. S. *Biochem. Biophys. Res. Comm.* **1990**, *171*, 445–450.

67. Harrison, R. K.; Stein, R. L. *Biochemistry* **1990**, *29*, 3813–3816.

68. Liu, J.; Albers, M. W.; Chen, C. M.; Schreiber, S. L.; Walsh, C. T. *Proc. Natl. Acad. Sci. USA* **1990**, *87*, 2304–2308.

69. Harrison, R. K.; Stein, R. L. *Biochemistry* **1990**, *29*, 1684–1689.

70. Rosen, M. K.; Standaert, R. F.; Galat, A.; Nakatsuka, M.; Schreiber, S. L. *Science* **1990**, *248*, 863–866.

71. Orozco, M.; Tirado-Rives, J.; Jorgensen, W. L. *Biochemistry* **1993**, *32*, 12864–12874.

72. Fischer, S.; Michnik, S.; Karplus, M. *Biochemistry* **1993**, *32*, 13830–13837.

73. Ikeda, Y.; Schultz, W.; Clardy, J.; Schreiber, S. L. *J. Am. Chem. Soc.* **1994**, *116*, 4143–4144.

74. Ke, H.; Mayrose, D.; Cao, W. *Proc. Natl. Acad. Sci. USA* **1993**, *90*, 3324–3328.

75. Kallen, J.; Walkinshaw, M. D. *FEBS Lett.* **1992**, *300*, 286–290.

76. London, R. E.; Davis, D. G.; Vavrek, R. J.; Stewart, J. M.; Handschumacher, R. E. *Biochemistry* **1990**, *29*, 10298–10302.

77. Zhao, Y.; Ke, H. *Biochemistry* **1996**, *35*, 7362–7368.

78. Konno, M.; Ito, M.; Hayano, T.; Takahashi, N. *J. Mol. Biol.* **1996**, *256*, 897–908.

79. Zhao, Y.; Ke, H. *Biochemistry* **1996**, *35*, 7356–7361.

80. Kakalis, L. T.; Armitage, I. M. *Biochemistry* **1994**, *33*, 1495–1501.

81. Zhao, Y.; Chen, Y.; Schutkowski, M.; Fischer, G.; Ke, H. *Structure* **1997**, *5*, 139–146.
82. Trandinh, C. C.; Pao, G. M.; Saier, M. H. *FASEB J.* **1992**, *6*, 3410–3420.
83. Nigam, S. K.; Jin, Y.-J.; Jin, M. J.; Bush, K. T.; Bierer, B. E.; Burakoff, S. J. *Biochem. J.* **1993**, *294*, 511–515.
84. Bush, K. T.; Hendrickson, B. A.; Nigam, S. K. *Biochem. J.* **1994**, *303*, 705–708.
85. Riviere, S.; Menez, Galat, A. *FEBS Lett.* **1993**, *315*, 247–251.
86. Jin, Y. J.; Burakoff, S. J. *Proc. Natl. Acad. Sci. USA* **1993**, *90*, 7769–7773.
87. Fretz, H.; Albers, M. A.; Galat, A.; Standaert, R. F.; Lane, W. S.; Burakoff, S. J.; Bierer, B. E.; Schreiber, S.L. *J. Am. Chem. Soc.* **1991**, *113*, 1409–1410.
88. Tai, P. K.-K.; Albers, M. A.; Chang, H.; Faber, L. E.; Schreiber, S. L. *Science* **1992**, *256*, 1315–1318.
89. Callebaut, L.; Renoir, J.-M.; Lebeau, M.-C.; Massol, N.; Burny, R.; Baulieu, E.-E.; Mornon, J.-P. *Proc. Natl. Acad. Sci. USA* **1992**, *89*, 6270–6274.
90. Chambraud, B.; Rouviere-Fourmy, N.; Radanyl, C.; Hsiao, K.; Peattie, D. A.; Livingston, D. J.; Baulieu, E.-E. *Biochem. Biophys. Res. Comm.* **1993**, *196*, 160–166.
91. Ruff, V. A.; Yem, A. W.; Munns, P. L.; Adams, L. D.; Reardon, I. M.; Deibel, M. R.; Leach, K. L. *J. Biol. Chem.* **1992**, *267*, 21285–21288.
92. Czar, M. J.; Owens-Grillo, J. K.; Yem, A. W.; Leach, K. L.; Deibel, M. R.; Welsh, M. J.; Pratt, W. B. *Mol. Endocrinol.* **1994**, *8*, 1731–1741.
93. Perrot-Applanat, M.; Cibert, C.; Geraud, G.; Renoir, J.-M.; Baulieu, E.-E. *J. Cell Sci.* **1995**, *108*, 2037–2051.
94. Zeng, B.; MacDonald, J. R.; Bann, J. G.; Beck, K.; Gambee, J. E.; Boswell, B. A.; Bachinger, H. P. *Biochem. J.* **1998**, *330*, 109–114.
95. Donnelly, J. G.; Soldin, S. J. *Clin. Biochem.* **1994**, *27*, 367–372.
96. Etzkorn, F. A.; Chang, Z. Y.; Stolz, L. A.; Walsh, C. T. *Biochemistry* **1994**, *33*, 2380–2388.
97. Spik, G.; Haendler, B.; Delmas, O.; Mariller, C.; Chamoux, M.; Maes, P.; Tartar, A.; Montreuil, J.; Stedman, K.; Kocher, H. P.; Keller, R.; Hiestand, P. C.; Movva, N. R. *J. Biol. Chem.* **1991**, *266*, 10735–10738.
98. Allain, F.; Boutillon, C.; Mariller, C.; Spik, G. *J. Immunol. Methods* **1995**, *178*, 113–120.
99. Friedman, J.: Trahey, M.; Weissman, I. *Proc. Natl. Acad. Sci. USA* **1993**, *90*, 6815–6819.
100. Bergsma, D. J.; Eder, C.; Gross, M.; Kerster, H.; Sylvester, D.; Appelbaum, E.; Cusimano, D.; Livi, G. P.; McLaughlin, M. M.; Kasyan, K.; Porter, T. G.; Silverman, C.; Dunnington, D.; Hand, A.; Prichett, W. P.; Bossard, M. J.; Brandt, M.; Levy, M. A. *J. Biol. Chem.* **1991**, *266*, 23204–23214.
101. Woodfield, K. Y.; Price, N. T.; Halestrap, A. P. *Biochim. Biophys. Acta* **1997**, *1351*, 27–30.
102. Tanveer, A.; Viriji, S.; Andreeva, L.; Totty, N. F.; Hsuan, J. J.; Ward, J. M.; Crompton, M. *Eur. J. Biochem.* **1996**, *238*, 166–172.
103. Hoffmann, K.; Kakalis, T.; Anderson, K. S.; Armitage, I. A.; Handschumacher, R. E. *Eur. J. Biochem.* **1995**, *229*, 188–193.
104. Kieffer, L.; Seng, T. W.; Li, W.; Osterman, D. G.; Handschumacher, R. E.; Bayney, R. M. *J. Biol. Chem.* **1993**, *268*, 12303–12310.

105. Ratajczak, T.; Carrello, A.; Mark, P. J.; Warner, B. J.; Simpson, R. J.; Moritz, R. L.; House, A. K. *J. Biol. Chem.* **1993**, *268*, 13187–13192.

106. Lilie, H.; Rudolph, R.; Buchner, J. *J. Mol. Biol.* **1995**, *248*, 190–201.

107. Freskgard, P.-O.; Bergenhem, N.; Jonsson, B.-H.; Svensson, M.; Carlson, Y. *Science* **1992**, *258*, 466–468.

108. Kiefhaber, T.; Quass, R.; Hahn, U.; Schmid, F. X. *Biochemistry* **1990**, *29*, 3061–3070.

109. Kern, G.; Drakenberg, T.; Wilkstrom, M.; Forsen, S.; Bang, H.; Fischer, G. *FEBS Lett.* **1993**, *323*, 198–202.

110. Kern, G.; Kern, D.; Schmid, F. X.; Fischer, G. *FEBS Lett.* **1994**, *348*, 145–148.

111. Tropschug, M.; Wachter, E.; Mayer, S.; Schonbrunner, E. R.; Scmid, F.X. *Nature* **1990**, *343*, 674–677.

112. Colley, N. J.; Baker, E. K.; Stamnes, M. A.; Zuker, C. S. *Cell* **1991**, *67*, 255–263.

113. Ondek, B.; Hardy, R. W.; Baker, E. K.; Stamnes, M. A.; Shieh, B.-H.; Zuker, C. S. *J. Biol. Chem.* **1992**, *267*, 16460–16466.

114. Lodish, H. F.; Kong, N. *J. Biol. Chem.* **1991**, *266*, 14835–14838.

115. Steinmann, B.; Bruckner, P. *J. Biol. Chem.* **1991**, *266*, 1299–1303.

116. Helekar, S. A.; Char, D.; Neff, S.; Patrick, J. *Neuron* **1994**, *12*, 179–189.

117. Davis, E. C.; Broekelmann, T. J.; Ozawa, Y.; Mecham, R. P. *J. Cell Biol.* **1998**, *140*, 295–303.

118. Callebaut, I.; Mornon, J. P. *FEBS Lett.* **1995**, *374*, 211–215.

119. Duina, A. A.; Chang, H.-C. J.; Marsh, J. A.; Lindquist, S.; Gaber, R.F. *Science* **1996**, *274*, 1713–1715.

120. Freeman, B. C.; Toft, D. O.; Morimoto, R. I. *Science* **1996**, *274*, 1718–1720.

121. Bose, S.; Weikl, T.; Bugl, H.; Buchner, J. *Science* **1996**, *274*, 1715–1718.

122. Owens-Grillo, J. K.; Hoffmann, K.; Hutchison, K. A.; Yem, A. W. *J. Biol. Chem.* **1995**, *270*, 20479–20484.

123. Coss, M. C.; Stephens, R. M.; Morrison, D. K.; Winterstein, D.; Smith, L. M.; Simek, S. L. *Cell Growth Differ.* **1998**, *9*, 41–48.

124. Uittenbogaard, A.; Ying, Y.; Smart, E. J. *J. Biol. Chem.* **1998**, *273*, 6525–6532.

125. Gardner, P. *Ann. Rev. Immun.* **1990**, *8*, 231–252.

126. Khan, A. A.; Steiner, J. P.; Klein, G.; Scneider, M. F.; Snyder, S. H. *Science* **1992**, *257*, 815–818.

127. Berridge, M. J.; Irvine, R. F. *Nature* **1989**, *341*, 197–204.

128. Ferris, C. D.; Snyder, S. H. *Ann. Rev. Physiol.* **1991**, *54*, 469–488.

129. Furuichi, T. *Nature* **1989**, *342*, 32–38.

130. Cameron, A. M.; Steiner, J. P.; Sabatini, D. M.; Kaplin, A. I.; Walensky, L. D.; Snyder, S. H. *Proc. Natl. Acad. Sci. USA* **1995**, *92*, 1784–1788.

131. Cameron, A. M.; Steiner, J. P.; Roskams, A. J.; Ali, S. M.; Ronnett, G. V.; Snyder, S.H. *Cell* **1995**, *83*, 463–472.

132. Cameron, A.M.; Nucifora, F.C.; Fung, E.T.; Livingston, D.J.; Aldape, R.A.; Ross, C. A.; Snyder, S. H. *J. Biol. Chem.* **1997**, *272*, 27582–27588.

133. Fleischer, S.; Inui, M. *Ann. Rev. Biophys. Biophys. Chem.* **1989**, *18*, 333–364.

134. Sorrentino, V.; Volpe, P. *Trends Pharm. Sci.* **1993**, *14*, 98–103.

135. Marks, A. R.; Tempst, P.; Hwang, K. S.; Taubman, M. B.; Inui, M.; Chadwick, C.; Fleischer, S.; Nadal-Ginard, B. *Proc. Natl. Acad. Sci. USA* **1989**, *86*, 8683–8687.

136. Takeshima, H.; Nishimura, S.; Matsumoto, T.; Ishida, H.; Kangawa, K.; Minamino, N.; Matsuo, H.; Ueda, M.; Hanaoka, M.; Hirose, T. et al. *Nature* **1989**, *339*, 439–445.

137. Otsu, K.; Willard, H. F.; Khanna, V. K.; Zorzato, F.; Green, N. M.; MacLennan, D. H. *J. Biol. Chem.* **1990**, *265*, 13472–13483.

138. Collins, J. H. *Biochem. Biophys. Res. Commun.* **1991**, *178*, 1288–1290.

139. Timerman, A. P.; Onoue, H.; Xin, H.-B.; Barg, S.; Copello, J.; Wiederrecht, G.; Fleischer, S. *J. Biol. Chem.* **1996**, *271*, 20385–20391.

140. Jayraman, T.; Brillantes, A.-M.; Timerman, A. P.; Fleischer, S.; Erdjument-Bromage, H.; Tempst, P.; Marks, A. R. *J. Biol. Chem.* **1992**, *267*, 9474–9477.

141. Brillantes, A.-M.; Ondrias, K.; Scott, A.; Kobrinsky, E.; Ondriasova, E.; Moschella, M. C.; Jayraman, T.; Landers, M.; Ehrlich, B. E.; Marks, A. R. *Cell* **1994**, *77*, 513–523.

142. Timerman, A. P.; Wiederrecht, G.; Marcy, A.; Fleischer, S. *J. Biol. Chem.* **1995**, *270*, 2451–2459.

143. Wagenknecht, T.; Radermacher, M.; Grassucci, R.; Berkowitz, J.; Xin, H.-B.; Fleischer, S. *J. Biol. Chem.* **1997**, *272*, 32463–32471.

144. Bram, R. J.; Crabtree, G. R. *Nature* **1994**, *371*, 355–358.

145. Halloway, M. P.; Bram, R. J. *J. Biol. Chem.* **1996**, *271*, 8549–8552.

146. Bram, R. J. *Gene* **1996**, *174*, 307–309.

147. Bang, H.; Muller, W.; Hans, M.; Brune, K.; Swandulla, D. *Proc. Natl. Acad. Sci. USA* **1995**, *92*, 3435–3438.

148. Sherry, B.; Yarlett, N.; Strupp, A.; Cerami, A. *Proc. Natl. Acad. Sci. USA* **1992**, *89*, 3511–3515.

149. Xu, Q.; Leiva, M. C.; Dischkoff, S. A.; Handschumacher, R. E.; Lyttle, C. R. *J. Biol. Chem.* **1992**, *267*, 11968–11971.

150. Endrich, M. M.; Grossenbacher, D.; Geistlach, A.; Gehring, H. *Biol. Cell* **1996**, *88*, 15–22.

151. Billich, A.; Winkler, G.; Aschauer, H.; Rot, A.; Peichl, P. *J. Exp. Med.* **1997**, *185*, 975–980.

152. Cirillo, R.; dePaulis, A.; Ciccarelli, A; Triggianai, M.; Marone, G. *Int. Arch. Allergy Appl. Immunol.* **1991**, *94*, 76–77.

153. Massague, J.; Attisano, L.; Wrana, J. L. *Trends Cell Biol.* **1994**, *4*, 172–178.

154. Wrana, J. L.; Attisano, L.; Weiser, R.; Ventura, F.; Massague, J. *Nature* **1994**, *370*, 341–347.

155. Wang, T. W.; Donahoe, P. K.; Zervos, A. S. *Science* **1994**, *265*, 674–676.

156. Wang, T.; Li, B.-Y.; Danielson, P. D.; Shah, P. C.; Rockwell, S.; Lechleider, R. J.; Martin, J.; Manganaro, T.; Donahoe, P. K. *Cell* **1996**, *86*, 435–444.

157. Yang, D.; Rosen, M. K.; Schreiber, S. L. *J. Am. Chem. Soc.* **1993**, *115*, 819–820.

158. Chang, M.-J.; Kinnunen, P.; Hawkers, J.; Brand, T.; Scneider, M. D. *J. Biol. Chem.* **1996**, *271*, 22941–22944.

159. Chen, Y.-G.; Liu, F.; Massague, F. *EMBO J.* **1997**, *16*, 3866–3876.

160. Okadome, T.; Oeda, E.; Saitoh, M.; Ichijo, H.; Moses, H. L.; Miyazono, K.; Kawabata, M. *J. Biol. Chem.* **1996**, *271*, 21687–21690.

161. Stockwell, B. R.; Schreiber, S. L. *Chem. Biol.* **1998**, *5*, 385–395.

162. Bush, K. T.; Henrickson, B. A.; Nigam, S. K. *Biochem. J.* **1994**, *303*, 705–708.

163. Partaledis, J. A.; Berlin, V. *Proc. Natl. Acad. Sci. USA* **1993**, *90*, 5450–5454.

164. Walensky, L. D.; Gascard, P.; Fields, M. E.; Blackshaw, S.; Conboy, J. G.; Mohandas, N.; Snyder, S. H. *J. Cell Biol.* **1998**, *141*, 143–153.
165. Koide, M.; Obata, K.; Iio, A.; Iida, M.; Harayama, H.; Yokota, M.; Tuan, R. S. *Heart Vessels* **1997**, *Suppl. 12*, 7–9.
166. Shou, W.; Aghdasi, B.; Armstrong, D. L.; Guo, Q.; Bao, S.; Charng, M.-J.; Mathews, L. M.; Schneider, M. D.; Hamilton, S. L.; Matzuk, M. M. *Nature* **1998**, *391*, 489–492.
167. Sananes, N.; Baulieu, E. E.; Le Goascogne, C. *Biol. Reprod.* **1998**, *58*, 353–360.
168. Montague, J. W.; Hughes, F. M.; Cidlowski, J. A. *J. Biol. Chem.* **1997**, *272*, 6677–6684.
169. Luban, J.; Bossolt, K.L.; Franke, E.K.; Kalpana, G.V.; Goff, S.P. *Cell* **1993**, *73*, 1067–1078.
170. Franke, E. K.; Yuan, H. E. H.; Luban, J. *Nature* **1994**, *372*, 359–362.
171. Thali, M.; Bukovsky, A.; Kondo, E.; Rosenwirth, B.; Walsh, C. T.; Sodroski, J.; Gottlinger, H.G. *Nature* **1994**, *372*, 363–365.
172. Papageorgiou, C.; Sanglier, J.-J.; Traber, R. *Bioorg. Med. Chem. Lett.* **1996**, *6*, 23–26.
173. Braaten, D.; Franke, E. K.; Luban, J. *J. Virol.* **1996**, *70*, 4220–4227.
174. Braaten, D.; Franke, E. K.; Luban, J. *J. Virol.* **1996**, *70*, 3551–3560.
175. Gamble, T. R.; Vajdos, F. F.; Yoo, S.; Worthylake, D. K.; Houseweart, M.; Sundquist, W. I.; Hill, C. P. *Cell* **1996**, *87*, 1285–1294.
176. Endrich, M. M.; Gehring, H. *Eur. J. Biochem.* **1998**, *252*, 441–446.
177. Braun, W.; Kallen, J.; Mikol, V.; Walkinshaw, M. D.; Wuthrich, K. *FASEB J.* **1995**, *9*, 63–72.
178. Sybyl reference
179. Michnick, S. W.; Rosen, M. K.; Wandless, T. J.; Karplus, M.; Schreiber, S. L. *Science* **1991**, *252*, 836–839.
180. Moore, J. M.; Peattie, D. A.; Fitzgibbon, M. J.; Thomson, J. A. *Nature* **1991**, *351*, 248–250.
181. (a) Chothia, C. *Ann. Rev. Biochem.* **1984**, *53*, 537–572. (b) Chirgadze, Y. N. *Acta Crystallogr.* **1987**, *A43*, 405–XX. (c) Ptitsyn, O. B.; Finkelstein, A. V. *Q. Rev. Biophys.* **1980**, *13*, 339–386.
182. Van Duyne, G. D.; Standaert, R. F.; Karplus, P. A.; Schreiber, S. L.; Clardy, J. *J. Mol. Biol.* **1993**, *229*, 105–124.
183. Van Duyne, G. D.; Standaert, R. F.; Karplus, P. A.; Schreiber, S. L.; Clardy, J. *Science* **1991**, *252*, 839–842.
184. Lepre, C. A.; Thomson, J. A.; Moore, J. M. *FEBS Lett.* **1992**, *302*, 89–96.
185. Meadows, R. P.; Nettesheim, D. G.; Xu, R. X.; Olejniczak, E. T.; Petros, A. M.; Holzman, T. F.; Severin, J.; Gubbins, E.; Smith, H.; Fesik, S.W. *Biochemistry* **1993**, *32*, 754–765.
186. Wilson, K. P.; Yamashita, M. M.; Sintchak, M. D.; Rotstein, S. H.; Murcko, M. A.; Boger, J.; Thomson, J. A.; Fitzgibbon, M. J.; Black, J. R.; Navia, M. A. *Acta Cryst.* **1995**, *D51*, 511–521.
187. Becker, J. W.; Rotonda, J.; McKeever, B. M.; Chan, H. K.; Marcy, A. I.; Wiederrecht, G.; Hermes, J. D.; Springer, J. P. *J. Biol. Chem.* **1993**, *268*, 11335–11339.

188. Holt, D. A.; Luengo, J. I.; Yamashita, D. S.; Oh, H.-J.; Konialian, A. L.; Yen, H.-K.; Rozamus, L. W.; Brandt, M.; Bossard, M. J.; Levy, M. A.; Eggleston, D. S.; Liang, J.; Schultz, L. W.; Stout, T. J.; Clardy, J. *J. Am. Chem. Soc.* **1993**, *115*, 9925–9938.

189. Armistead, D. M.; Badia, M. C.; Deininger, D. D.; Duffy, J. P.; Saunders, J. O.; Tung, R. D.; Thomson, J. A.; DeCenzo, M. T.; Futer, O.; Livingston, D. J.; Murcko, M. A.; Yamashita, M. M.; Navia, M. A. *Acta Cryst.* **1995**, *D51*, 522–528.

190. Dragovich, P. S.; Barker, J. E.; French, J.; Imbacuan, M.; Kalish, V. J.; Kissinger, C. R.; Knighton, D. R.; Lewis, C. T.; Moomaw, E. W.; Parge, H. E.; Pelletier, L. A. K.; Prins, T. J.; Showalter, R. E.; Tatlock, J. H.; Tucker, K. D.; Villafranca, J. E. *J. Med. Chem.* **1996**, *39*, 1872–1884.

191. Wiederrecht, G.; Hung, S.; Chan, H. K.; Marcy, A.; Martin, M.; Calaycay, J.; Boulton, D.; Sigal, N.; Kincaid, R. L.; Siekerka, J. J. *J. Biol. Chem.* **1992**, *267*, 21753–21760.

192. Bossard, M. J.; Bergsma, D. J.; Brandt, M.; Livi, G. P.; Eng, W.-K.; Johnson, R. K.; Levy, M. A. *Biochem. J.* **1994**, *297*, 365–372.

193. DeCenzo, M. T.; Park, S. T.; Jarrett, B. P.; Aldape, R. A.; Futer, O.; Murcko, M. A.; Livingston, D. J. *Protein Eng.* **1996**, *9*, 173–180.

194. Aldape, R. A.; Futer, O.; DeCenzo, M. T.; Jarrett, B. P.; Murcko, M. A.; Livingston, D. J. *J. Biol. Chem.* **1992**, *267*, 16029–16032.

195. Griffin, J. P.; Kim, J. L.; Kim, E. E.; Sintchak, M. D.; Thomson, J. A.; Fitzgibbon, M. J.; Fleming, M. A.; Caron, P. R.; Hsiao, K.; Navia, M. A. *Cell* **1995**, *82*, 507–522.

196. Kissinger, C. R.; Parge, H. E.; Knighton, D. R.; Lewis, C. T.; Pelletier, L. A.; Tempczyk, A.; Kalish, V. J.; Tucker, K. D.; Showalter, R. E.; Moomaw, E. W.; Gastinel, L. N.; Habuka, N.; Chen, X.; Maldonado, F.; Barker, J. E.; Bacquet, R.; Villafranca, J. E. *Nature* **1995**, *378*, 641–644.

197. Choi, J.; Chen, J.; Schreiber, S. L.; Clardy, J. *Science* **1996**, *273*, 239–242.

198. Schultz, L. W.; Martin, P. K.; Liang, J.; Schreiber, S. L.; Clardy, J. *J. Am. Chem. Soc.* **1994**, *116*, 3129–3130.

199. Liang, J.; Hung, D. T.; Schreiber, S. L.; Clardy, J. *J. Am. Chem. Soc.* **1996**, *118*, 1231–1232.

200. Craescu, C. T.; Rouviere, N.; Popescu, A.; Cerpolini, E.; Lebeau, M.-C.; Baulieu, E.-E.; Misplelter, J. *Biochemistry* **1996**, *35*, 11045–11052.

201. Ke, H.; Zydowsky, L. D.; Liu, J.; Walsh, C. T. *Proc. Natl. Acad. Sci. USA* **1991**, *88*, 9483–9487.

202. Ke, H. *J. Mol. Biol.* **1992**, *228*, 539–550.

203. Kallen, J.; Spitzfaden, C.; Zurini, M. G. M .; Wider, G.; Widmer, H.; Wuthrich, K.; Walkinshaw, M. D. *Nature* **1991**, *353*, 276–279.

204. Weber, C.; Wider, G.; Von Freyburg, B.; Traber, R.; Braun, W.; Widmer, H.; Wuthrich, K. *Biochemistry* **1991**, *30*, 6563–6574.

205. Theriault, Y.; Logan, T. M.; Meadows, R.; Yu, L.; Olejniczak, E. T.; Holzman, T. F.; Simmer, R. L.; Fesik, S. W. *Nature* **1993**, *361*, 88–91.

206. Pfluegl, G.; Kallen, J.; Schirmer, T.; Jansonius, J. N.; Zurini, M. G.; Walkinshaw, M. D. *Nature* **1993**, *361*, 91–94.

207. Mikol, V.; Kallen, J.; Pfluegl, G.; Walkinshaw, M. D. *J. Mol. Biol.* **1993**, *234*, 1119–1130.

208. Spitzfaden, C.; Braun, W.; Wider, G.; Widmer, H.; Wuthrich, K. *J. Biomol. NMR* **1994**, *4*, 463–482.
209. Liu, J.; Chen, C. M.; Walsh, C. T. *Biochemistry* **1991**, *30*, 23206–2310.
210. Bossard, M. J.; Koser, P. L.; Brandt, M.; Bergsma, D. J. *Biochem Biophys. Res. Comm.* **1991**, *176*, 1141–1148.
211. Zydowsky, L. D.; Etzkorn, F. A.; Chang, H. Y.; Ferguson, S. B.; Stolz, L. A.; Ho, S. I.; Walsh, C. T. *Protein Sci.* **1992**, *9*, 1092–1099.
212. Mikol, V.; Kallen, J.; Walkinshaw, M. D. *Proc. Natl. Acad. Sci. USA* **1994**, *91*, 5183–5186.
213. Ke, H.; Zhao, Y.; Luo, F.; Weissman, I.; Friedman, J. *Proc. Natl. Acad. Sci. USA* **1993**, *90*, 11850–11854.
214. Babine, R. E.; Bender, S. L. *Chem. Rev.* **1997**, *97*, 1359–1472.
215. Veerapandian, P. (Ed.). *Structure-Based Drug Design*; Marcel Dekker: New York, 1997.
216. Charifson, P. S. (Ed.). *Practical Application of Computer-Aided Drug Design*; Marcel Dekker: New York, 1997.
217. Greer, J.; Erickson, J. W.; Baldwin, J. J.; Varney, M. D. *J. Med. Chem.* **1994**, *37*, 1035–1054.
218. Pascard, C. *Acta. Cryst.* **1995**, *D51*, 407–417.
219. Erickson, J. W.; Fesik, S. W. *Ann. Rep. Med. Chem.* **1992**, *27*, 271–289.
220. Archer, S. J.; Domaille, P. J. *Ann. Rep. Med. Chem.* **1996**, *31*, 299–308.
221. Holt, D. A.; Konialian-Beck, A. L.; Oh, H.-J.; Yen, H.-K.; Rozamus, L. W.; Krog, A. J.; Erhard, K. F.; Ortiz, E.; Levy, M. A.; Brandt, M.; Bossard, M. J.; Luengo, J. I. *BioMed. Chem. Lett.* **1994**, *4*, 315–320.
222. Hamilton, G. S.; Steiner, J. P. *Curr. Pharm. Design* **1997**, *3*, 405–428.
223. Lamb, M. L.; Jorgensen, W. L. *J. Med. Chem.* **1998**, *41*, 3928–3939.
224. LeMaster, D. M. *Progress NMR Spectrosc.* **1994**, *26*, 371–419.
225. Sattler, M.; Fesik, S. W. *J. Am. Chem. Soc.* **1997**, *119*, 33, 7885–7886.
226. Hadjuk, P. J.; Olejniczak, E. T.; Fesik, S. W. *J. Am. Chem. Soc.* **1997**, *119*, 12257–12261.
227. Shuker, S. B.; Hadjuk, P. J.; Meadows, R. P.; Fesik, S. W. *Science* **1996**, *274*, 1531–1534.
228. Hadjuk, P. J.; Sheppard, G.; Nettesheim, D. G.; Olejniczak, E.T.; Shuker, S. B.; Meadows, R. P.; Steinman, D. H.; Carrera, G.M.; Marcotte, P. A.; Severin, J.; Walter, K.; Smith, H.; Gubbins, E.; Simmer, R.; Holzman, T. F.; Morgan, D. W.; Davidsen, S. K.; Summers, J. B.; Fesik, S. W. *J. Am. Chem. Soc.* **1997**, *119*, 5818–5827.
229. Cheng, J.-W.; Lepre, C. A.; Chambers, S. P.; Fulghum, J. R.; Thomson, J. A.; Moore, J. M. *Biochemistry* **1993**, *32*, 9000–9010.
230. Lepre, C. A.; Pearlman, D. A.; Cheng, J.-W.; DeCenzo, M. T.; Livingston, D. J.; Moore, J. M. *Biochemistry* **1994**, *33*, 13571–13580.
231. Itoh, S.; DeCenzo, M. T.; Livingston, D. J.; Pearlman, D. A.; Navia, M. A. *Bioorg. Med. Chem. Lett.* **1995**, *5*, 1983–1988.
232. Ivery, M. T.; Weiler, L. *Bioorg. Med. Chem.* **1997**, *5*, 217–232.
233. Lepre, C. A.; Pearlman, D. A.; Futer, O.; Livingston, D. J.; Moore, J. M. *J. Biomolec. NMR* **1996**, *8*, 67–76.

234. Futer, O.; DeCenzo, M. T.; Park, S.; Jarrett, B.; Aldape, R.; Livingston, D. J. *J. Biol. Chem.* **1995**, *270*, 18935–18940.

235. Rosen, M. K.; Yang, D.; Martin, P. K.; Schreiber, S. L. *J. Am. Chem. Soc.* **1993**, *115*, 821–822.

236. The reader is also referred to the following reviews of this topic: (a) Snyder, S. H.; Sabatini, D. M. *Nature Medicine* **1995**, *1*, 32–37. (b) Snyder, S. H.; Sabatini, D. M.; Lai, M. M.; Steiner, J. P.; Hamilton, G. S.; Suzdak, P. D. *Trends. Pharm. Sci.* **1998**, *19*, 21–26. (c) Hamilton, G. S.; Steiner, J. P. *J. Med. Chem.* **1999**, in press.

237. Maki, N.; Sekiguchi, F.; Nishimaki, J.; Miwa, K.; Hayano, T.; Takahashi, N.; Suzuki, M. *Proc. Natl. Acad. Sci. USA* **1990**, *87*, 5440–5443.

238. Steiner, J. P.; Dawson, T. M.; Fotuhi, M.; Glatt, C. E.; Snowman, A. M.; Cohen, N.; Snyder, S. H. *Nature* **1992**, *358*, 584–587.

239. Dawson, T. M.; Steiner, J. P.; Lyons, W. E.; Fotuhi, M.; Blue, M.; Snyder, S. H. *Neuroscience* **1994**, *62*, 569–580.

240. Dawson, T. M.; Steiner, J.P.; Dawson, V.L.; Dinerman, J.L.; Uhl, G.R.; Snyder, S. H. *Proc. Natl. Acad. Sci. USA* **1993**, *90*, 9808–9812.

241. Palmer, R. M.; Higgs, E.A. *Pharmacol. Rev.* **1991**, *43*, 109–142.

242. Dawson, T. M.; Snyder, S. H. *J. Neurosci.* **1994**, *14*, 5147–5159.

243. Bredt, D. S.; Ferris, C. D.; Snyder, S. H. *J. Biol. Chem.* **1992**, *267*, 10976–10981.

244. Bredt, D. S.; Snyder, S. H. *Proc. Natl. Acad. Sci. USA* **1989**, *86*, 9030–9033.

245. Choi, D. W. *Science* **1992**, *258*, 241–243.

246. Dawson, V. L.; Dawson, T. M.; London, E. D.; Bredt, D. S.; Snyder, S. H. *Proc. Natl. Acad. Sci. USA* **1991**, *88*, 6368–6371.

247. Dawson, V. L.; Kizushi, V. M.; Huang, P. L.; Snyder, S. H.; Dawson, T. M. *J. Neurosci.* **1996**, *16*, 2479–2487.

248. Sharkey, J.; Butcher, S. P. *Nature* **1994**, *371*, 336–339.

249. Tokime, T.; Nozaki, K.; Kikuchi, H. *Neurosci. Lett.* **1996**, *206*, 81–84.

250. Ide, T.; Morikawa, E.; Kirino, T. *Neurosci. Lett.* **1996**, *204*, 157–160.

251. Butcher, S. P.; Henshall, D. C.; Teramura, Y.; Iwasaki, K.; Sharkey, J. *J. Neurosci.* **1997**, *17*, 6939–6946.

252. Yagita, Y.; Kitagawa, K.; Matsushita, K.; Taguchi, A.; Mabuchi, T.; Ohtsuki, T.; Yanagihara, T.; Matsumoto, M. *Life Sci.* **1996**, *59*, 1643–1650.

253. Steiner, J. P.; Dawson, T. M.; Fotuhi, M.; Snyder, S. H. *Mol. Med.* **1996**, *2*, 325–333.

254. Nichols, R. A.; Suplick, G. R.; Brown, J. M. *J. Biol. Chem.* **1994**, *269*, 23817–23823.

255. Hirsch, D. B. *Curr. Biol.* **1993**, *3*, 749–754.

256. Meiri, K. F.; Bickerstaff, L. E.; Schwob, J. E. *J. Cell Biol.* **1991**, *112*, 991–1005.

257. Lyons, W. E.; Steiner, J. P.; Snyder, S. H.; Dawson, T. M. *J. Neurosci.* **1995**, *15*, 2985–2994.

258. Lyons, W. E.; George, E. B.; Dawson, T. M.; Steiner, J. P.; Snyder, S. H. *Proc. Natl. Acad. Sci. USA* **1994**, *91*, 3191–3195.

259. Gold, B. G. *Mol. Neurobiol.* **1997**, *15*, 285–306.

260. Chang, H. Y.; Takei, K.; Sydor, A. M.; Born, T.; Rusnak, F.; Jay, D. G. *Nature* **1995**, *376*, 686–690.

261. Gold, B.G.; Storm-Dickerson, T.; Austin, D.R. *Restorative Neurol. Neurosci.* **1994**, *6*, 287.

262. Gold, B. G.; Storm-Dickerson, T.; Austin, D. R. *J. Neurosci.* **1995**, *15*, 7509–7516.

263. Steiner, J. P.; Connolly, M. A.; Valentine, H. L.; Hamilton, G. S.; Dawson, T. M.; Hester, L.; Snyder, S. H. *Nature Med.* **1997**, *3*, 421–428.

264. Hamilton, G. S.; Huang, W.; Connolly, M. A.; Ross, D. T.; Guo, H.; Valentine, H. L.; Suzdak, P. D.; Steiner, J. P. *Bioorg. Med. Chem. Lett.* **1997**, *7*, 1785–1790.

265. Li, J.-H.; Hamilton, G. S.; Huang, W.; Connolly, M. A.; Ross, D. T.; Guo, H.; Valentine, H. L.; Steiner, J. P. *214th National American Chemical Society Meeting*, 1997, Abstract MEDI 183.

266. Steiner, J. P.; Hamilton, G. S.; Ross, D. T.; Valentine, H. L.; Guo, H.; Connolly, M. A.; Liang, S.; Ramsey, C.; Li, J.-H.; Huang, W.; Howorth, P.; Soni, R.; Fuller, M.; Sauer, H.; Nowotnick, A.; Suzdak, P. D. *Proc. Natl. Acad. Sci. USA* **1997**, *94*, 2019–2024.

267. Woolf, C. J.; Ma, Q. P.; Allchorne, A.; Poole, S. J. *J. Neurosci.* **1996**, *16*, 2716–2723.

268. Gold, B. G.; Zeleney,-Pooley, M.; Wang, M. S.; Chaturvedi, P.; Armistead, D. M. *Exp. Neurobiol.* **1997**, *147*: 269–278.

269. Gold, B. G.; Zeleney-Poole, M.; Wang, M.-S.; Chaturvedi, P.; Armistead, D. M.; McCaffrey, P. G. *Soc. Neurosci. Abstr.* **1997**, 449.12.

270. Valentine, H. L.; Spicer, D.; Fuller, M.; Hamilton, G. S.; Steiner, J. P. *Soc. Neurosci. Abstr.* **1998**, 185.9

271. Liang, S.; Valentine, H. L.; Ramsey, C.; Soni, R.; Scott, C.; Steiner, J. P. *Soc. Neurosci. Abstr.* **1998**, 185.9

272. Hamilton, G. S. *Chem. Ind.* **1998**, Feb. 16, 127–132.

273. Gerlach, M.; Riederer, P.; Przuntek, H.; Youdim, M. B. *Eur. J. Pharmacol.* **1991**, *208*, 273–286.

274. Steiner, J. P.; Ross, D. T.; Valentine, H. L.; Liang, S.; Guo, H.; Huang, W.; Suzdak, P. D; Hamilton, G. S. *Soc. Neurosci. Abstr.* **1997**, 677.6.

275. Steiner, J. P.; Ross, D. T.; Sauer, H.; Hamilton, G. S., Valentine, H. L.; Guo, H.; Liang, S.; Spicer, D.; Howorth, P.; Chen, Y.; Fuller, M.; Suzdak, P. D. *J. Neurosci.* **1999**. In press.

276. Sachs, G.; Jonsson, G. *Biochem. Pharmacol.* **1975**, *24*, 1–8.

277. Ungerstedt, U.; Aburthnott, G. W. *Brain Res.* **1970**, *24*, 485–493.

278. Ross, D. T.; Guo, H.; Howorth, P.; Fuller, M.; Nowotnick, A.; Hamilton, G. S.; Suzdak, P. D.; Steiner, J. P. *Soc. Neurosci. Abstr.* **1997**, 677.7.

279. Guo, H.; Spicer, D. M.; Howorth, P.; Hamilton, G. S.; Suzdak, P. D; Ross, D. T. *Soc. Neurosci. Abstr.* **1997**, 677.12.

280. Bartus, R.T.; Dean, R.L.; Beer, B.; Lippa, A. S. *Science* **1992**, *217*, 408–417.

281. Sauer, H.; Francis, J. M.; Jiang, H.; Steiner, J. P. *Soc. Neurosci. Abstr.* **1998**, 269.9.

282. Sauer, H.; Francis, J. M.; Jiang, H.; Steiner, J. P. *Neuroscience* **1999**. In press.

283. Liang, S.; Valentine, H. L.; Ross, D. T.; Huang, W.; Suzdak, P. D.; Hamilton, G. S.; Steiner, J. P. *Soc. Neurosci. Abstr.* **1997**, 677.10.

284. Steiner, J. P.; Ho, T.; Coccia, C.; Griffin, J.; Snyder, S.; Rothstein, J. D. *Soc. Neurosci. Abstr.* **1998**, 121.1.

285. Ross, D. T.; Guo, H.; Chen, Y.; Howorth, P.; Steiner, J. P. *Soc. Neurosci. Abstr.* **1998**, 324.6.

286. Pawson, T. *Nature* **1995**, *373*, 573–580.

287. Coghlan, V. M.; Perrino, B. A.; Howard, M.; Langeberg, L. K.; Hicks, J. B.; Gallatin, W. M.; Scott, J.D. *Science* **1995**, *267*, 108–111.

288. Woolf, C. J.; Ma, Q. P.; Allchorne, A.; Poole, S. J. *J. Neurosci.* **1996**, *16*, 2716–2723.

289. Wang, M.S.; Zeleney-Poole, M.; Gold, B.G. *J. Pharm. Exp. Ther.* **1997**, *282*, 1084–1093.

290. Boriongan, C.V.; Stahl, C.E.; Fujisaki, T.; Porter, L.L.; Wtanabe, S. *Soc. Neurosci. Abstr.* **1998**, 778.16

291. Goto, S.; Singer, W. *Cereb. Cortex* **1994**, *4*, 636–645.

DISCOVERY OF ABT-594
AND RELATED NEURONAL
NICOTINIC ACETYLCHOLINE
RECEPTOR MODULATORS AS
ANALGESIC AGENTS:
MEDICINAL CHEMISTRY AND BIOLOGY

Mark W. Holladay and Michael W. Decker

Advances in Medicinal Chemistry
Volume 5, pages 85–113.
Copyright © 2000 by JAI Press Inc.
All rights of reproduction in any form reserved.
ISBN: 0-7623-0593-2

ABSTRACT

ABT-594, a nicotinic acetylcholine receptor (nAChR) modulator that exhibits potent antinociceptive activity in animal models of pain, was discovered through optimization of a series of compounds that was first identified as part of a program directed toward the discovery of nAChR modulators for Alzheimer's disease. Structure–activity studies on ABT-594 indicate that both the azetidine ring and an appropriately substituted pyridine ring are key structural features contributing to its biological activity, which together with its favorable pharmacokinetic and safety profiles, has led to its advancement to clinical trials for treatment of pain in humans.

I. INTRODUCTION

ABT-594 (**1**) is a nicotinic acetylcholine receptor (nAChR) modulator in clinical development for the treatment of pain. ABT-594 shows efficacy similar to that of morphine in several pain models and represents the first attempt to develop an analgesic clinical candidate using nAChR modulation as the mechanism of action. These facts together with its connection to a rare South American frog have resulted in a significant level of interest in ABT-594 in the scientific and lay communities. ABT-594 is a member of the 3-pyridyl ether class of compounds, a series which was first discovered during efforts to exploit nAChR modulators as potential treatments for Alzheimer's disease. In this chapter, the sequence of events leading to the discovery and characterization of ABT-594 will be described, followed by a summary of the properties of this and related compounds in a variety of biological systems that are relevant to their potential use as agents for the treatment of pain.

II. BACKGROUND: SUBTYPES OF NICOTINIC ACETYLCHOLINE RECEPTORS

The nAChR in skeletal muscle has been known for many years and has been extensively studied.[1-3] It is composed of five protein subunits [two α, and one each of β, γ (or ε), and δ] arranged around a central pore that forms an ion channel. Agonist binding results in channel opening and ion flux through the cell membrane. A pharmacologically distinct nAChR subtype in autonomic ganglia also has been known for many years. Molecular biology techniques have led to identification of mRNA for numerous additional nAChR subunits in neuronal tissue (α2–α8 and β2–β4), as well as α9 from rat cochlea. Heterologous expression studies have shown that numerous pairwise and/or triplex combinations of α2–α6 and β2–β4 subunits form functioning ion channels, whereas only α7, α8, and α9 can form homomeric channels. Thus many different subtypes of nAChRs are theoretically possible. Efforts to elucidate which combinations exist in nature and what are their functional roles continue to be subjects of intense investigation. It is now believed that the nAChR subtype in autonomic ganglia consists of a group of several related subtypes containing the α3 subunit in various combinations with α5, β2, and β4. In brain, two major subtypes are known. The α4β2 subtype is widely distributed and is labeled with high affinity by [³H]nicotine, [³H]cytisine, and numerous other classical nicotinic alkaloids. The α7 subtype binds nicotine with low affinity and α-bungarotoxin with high affinity, and exhibits a different pattern of distribution in the CNS than α4β2. The native α7 nAChR is probably identical to the homopentameric α7 nAChR observed in heterologous expression studies. The diverse pharmacology of behavioral and neurochemical responses to various

Table 1. Summary of Major Known Endogenous nAChR Subtypes

Localization	Subunits	Function
Skeletal Muscle	α1 β1 γ δ (ε)	Motor activity
Autonomic ganglia	α3 + α5/β2/β4	Autonomic neurotransmission
Brain	α4β2	Neurotransmitter release
Brain	α7	Glutamate release, Ca²⁺ influx
Brain, spinal cord, sensory ganglia, other locations	Additional combinations	Neurotransmitter release, sensory neurotransmission

nAChR modulators supports the existence of additional nAChR subtypes in the brain and spinal cord that are not yet fully characterized.[4] The major known endogenous nAChR subtypes are summarized in Table 1.

III. HISTORICAL PERSPECTIVE: EARLY WORK ON nAChR MODULATORS AT ABBOTT LABORATORIES

In 1989, Mike Williams joined the Neuroscience Discovery group at Abbott with the challenge of refocusing research activities toward a more aggressive drug discovery mode. His initial effort involved redirection of an existing effort in Alzheimer's disease that was, like many others, focused on muscarinic agonists as a palliative therapy. Many other companies were targeting the same approach, and compounds active at muscarinic receptors are highly prone to unacceptable side-effect liabilities. Therefore, it was felt that Abbott could be more competitive by focusing in the area of nicotinic receptor agonists, whereby two acute, albeit limited, human trials had shown beneficial action of nicotine in improving cognitive function in Alzheimer's patients. The decision was somewhat risky, inasmuch as there were few programs of this type in the industry, inevitably raising questions regarding what others knew that we did not, and vice versa. Moreover, the word nicotine immediately conjured associations with tobacco. However, it was reasoned that compounds selective for nAChR subtypes would have the potential for targeting the beneficial actions of nicotine while reducing or even eliminating its side effects.

The nascent nAChR project was led by Steve Arneric, a pharmacologist with a strong background in cardiovascular and central nervous system physiology. To put the concerns about association with nicotine and tobacco into a scientifically valid perspective, the term *cholinergic channel activator* (ChCA) was coined. The rationale behind this can be illustrated by analogy with the interaction of the hallucinogen LSD with serotonin (5HT) receptors. If 5HT receptors had been named LSD receptors based on the actions of this exogenous ligand, then there would have been the invalid assumption that all ligands interacting with 5HT receptors were LSD-like and sure to share the hallucinogenic properties of LSD. Instead, 5HT ligands have proven to be a rich source of new drugs as receptor subtype selective compounds have been developed.

Among several approaches undertaken by the medicinal chemistry team, at that time under the direction of David Garvey, toward discovering novel ChCAs for Alzheimer's disease, one based on structural modifications of nicotine itself eventually proved to be promising. A number of heterocycles had previously been shown to serve as mimics for the acetoxy group of acetylcholine in studies on the muscarinic system.[5,6] On the basis of this precedent, the pyridine ring of nicotine (2) was replaced with other heterocyclic rings—for example, isoxazole and isothiazole, leading to ABT-418 (3).[7,8] The biological profile of ABT-418 was generally similar to that of nicotine; however, a greater degree of separation between desirable and undesirable effects compared to nicotine could be demonstrated, and thus ABT-418 offered the potential to improve on nicotine as a therapeutic agent.[9] In the paper describing structure–activity relationships in the isoxazole series,[8] the analogue (4) lacking the 3-methyl group on the isoxazole is not included, since efforts to synthesize it had been unsuccessful up to that time. Eventually, compound 4 was prepared, and was shown to possess about 50-fold lower affinity than ABT-418 for the [³H]cytisine binding site (unpublished data). How fortunate that this compound was not targeted first as the representative isoxazole modification, since there is no obvious rationale to account for the greater potency afforded by the 3-methyl group, which is lacking in nicotine itself.

In 1993, as ABT-418 progressed through the course of early clinical trials, the pressing mission of the medicinal chemistry group was to prepare potential backup compounds. A matter of some debate was whether another isoxazole-like compound should suffice as a backup, or whether an entirely different series needed to be identified. A number of known nAChR ligands with diverse structures, such as anatoxin-a (5), 1,1-dimethyl-4-phenylpiperazinium (6), and *N*-methylcarbamyl choline (7), could potentially serve as lead compounds, and indeed many of these, as well as numerous isoxazole variants, were explored to at least some degree.

ABT-594 (1)

(S)-Nicotine (2)

ABT-418 (3)

des-3-methyl-ABT-418 (4)

Anatoxin-a (5)

1,1-Dimethyl-4-phenyl-piperazinium (6)

N-Methylcarbamylcholine (7)

IV. THE PYRIDYL ETHER SERIES

Meanwhile, compounds prepared previously during the muscarinic program were screened for nicotinic activity. This approach yielded notable success when compound **8** (Figure 1), a pyridazinyl ether, was found to have affinity for the rat brain [³H]cytisine nAChR binding site in the nanomolar range (K_i = 46 nM, Table 2).[10] Prepared in 1989 by John Chung, compound **8** had been found to be a weak ligand for muscarinic receptors (K_i = >10,000 nM at both M_1 and M_2 subtypes). Melwyn Abreo, noting the presence of the 2-pyrrolidinyl moiety in nicotine, prepared the corresponding 2-pyrrolidinyl analogue **9** (Figure 2), which was shown to possess similar affinity to that of **8** (K_i = 29 nM). Borrowing again from the structure of nicotine, the heteroaryl moieties of **8** and **9** were replaced with 3-pyridyl (Figure 3), yielding compounds **10** and **11**, respectively. Whereas compound **10** showed an impressive 10-fold improvement in binding affinity (K_i = 3 nM), the impact of this result was overshadowed by the enthusiasm generated by compound **11**, which possessed affinity for the [³H]cytisine binding site in the picomolar range (K_i = 0.15 nM).[11] Thus was born the prototype 3-pyridyl ether, A-84543 (**11**), which at the time was the highest affinity ligand known for the nAChR binding sites labeled by [³H]cytisine.

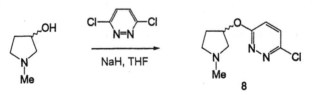

Figure 1. Synthesis of a pyridazinyl ether in the 3-substituted pyrrolidine series.

Table 2. nAChR Affinity of Compounds
8–11 and (*S*)-Nicotine

Compound	K_i (nM)[a]
(*S*)-nicotine (**2**)	1.0
8	46
9	29
10	3
A-84543 (**11**)	0.15

Note: [a]Displacement of [^{3}H]cytisine from rat brain membranes.

The subsequent structure–activity studies on A-84543 focused initially on the effects of stereochemistry, N-demethylation, azacycle ring size, and additional heteroaryl moieties. It was known that the (*R*)-enantiomer of nicotine and both stereoisomers of nornicotine were active in nicotinic assays, although generally weaker than (*S*)-nicotine.[8] It also was known that replacement of the pyrrolidine ring of (*S*)-nicotine with azetidine yielded a compound with equivalent[12] or higher[13] affinity than nicotine, whereas the piperidine compound was weaker.[13] The ready availability of (*S*)- and (*R*)-proline and (*S*)-azetidinecarboxylic acid as starting materials, together with the facile ether-forming chemistry described in Figure 3, permitted initial rapid development of the SAR shown in Table 3.[11,14] Interestingly, the structure–activity pattern in the pyridyl ether series with respect to stereochemistry and N-alkylation differs from that of nicotine (Figure 4).

The divergent SAR illustrated by Figure 4 raises the issue whether **11** may be interacting with the receptor site in a different mode than does nicotine. Is it even possible for low-energy conformations of **11** to reasonably superimpose on low energy conformations of nicotine? Fig-

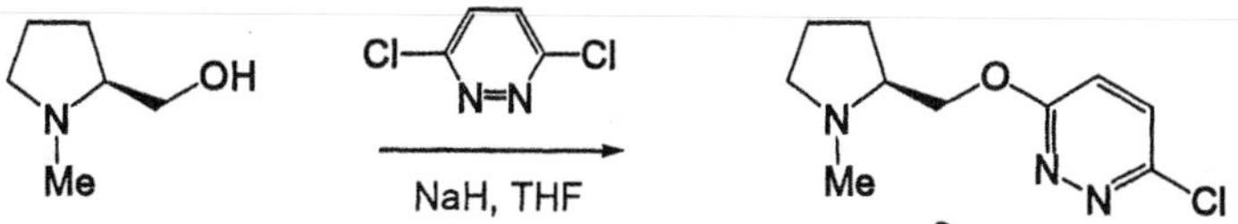

Figure 2. Synthesis of a pyridazinyl ether in the 2-substituted pyrrolidine series.

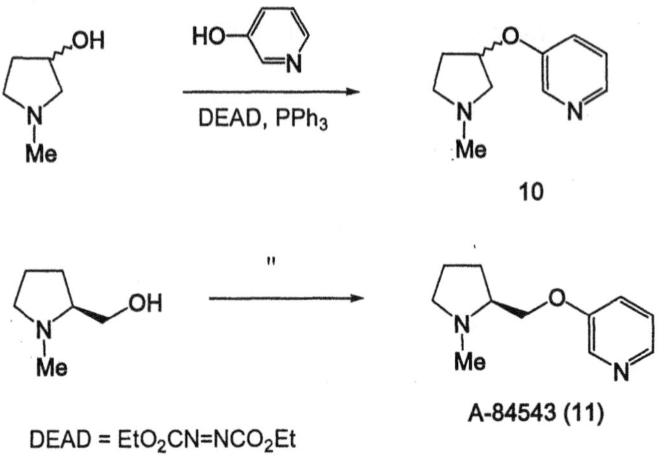

10

A-84543 (11)

DEAD = EtO$_2$CN=NCO$_2$Et

Figure 3. Synthesis of pyridyl ethers in the 3- and 2-substituted pyrrolidine series.

ure 5 illustrates a possible superposition of the two molecules (alternative possibilities have been discussed in ref. 11). One possible concern with the superposition suggested in Figure 5 is whether the receptor could accommodate the extra space occupied by the pyrrolidine ring of **11** (see Figure 5). The potent activity (unpublished data) of compound **18** (Figure

Table 3. nAChR Affinities of Analogues of A-84543

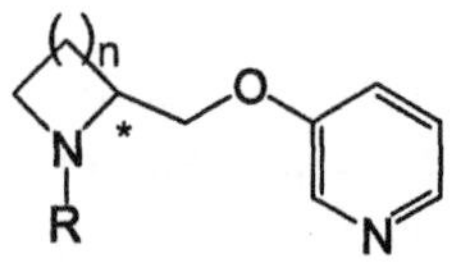

Compound	n	Stereochemistry	R	K$_i$ (nM)[a]
A-84543 (11)	2	S	Me	0.15
12	2	R	Me	19.7
13	2	S	H	0.16
14	2	R	H	0.14
15	1	S	Me	0.45
16	1	S	H	0.05
17	3	S	Me	73

Note: [a]Displacement of [^{3}H]cytisine from rat brain membranes.

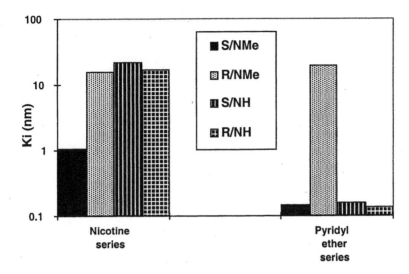

Figure 4. [³H]Cytisine binding affinities of (*S*)-nicotine (S/NMe), (*R*)-nicotine (R/NMe), (*S*)-nornicotine (S/NH), and (*R*)-nornicotine (R/NH) compared with the corresponding analogues (**11–14**, Table 3) in the pyridyl ether series. Data from refs. 7 and 11.

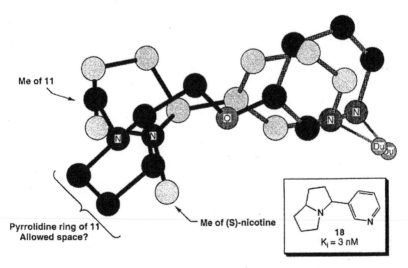

Figure 5. A possible superposition of pyridyl ether **11** (dark carbon atoms) and (*S*)-nicotine (light carbon atoms). "Du" represents a point on the receptor that could be interacting with the pyridine lone pair. Conformations were analyzed and overlays were performed using Chem 3D®.

Table 4. nAChR Affinities of Pyridine Ring Replacements in A-84543 (**11**)

Compound	Het	K_i (nM)[a]
11		0.15
19		495
20		7914
21		42
22		16
23		2
24		42[b]
25		4.7[b]
26		1.5[c]
27		2400[c]

Notes: [a]Displacement of [³H]cytisine from rat brain membranes; data from ref. 10 unless otherwise indicated.
[b]Data from ref. 15.
[c]Unpublished data.

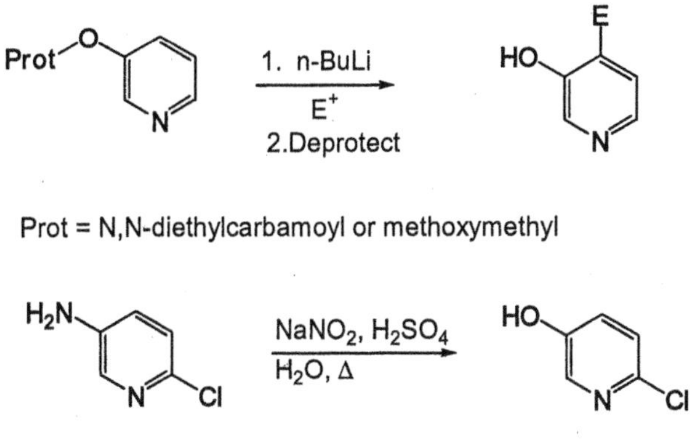

Prot = N,N-diethylcarbamoyl or methoxymethyl

Figure 6. Synthesis of some substituted 3-pyridinols.[16]

5) supports the notion that the answer to this question could be yes, since the distal azacyclic ring of a low energy conformer of **18** was found to fit well into the region in question.

Other structure–activity studies based on **11** examined additional variants of the heteroaryl moiety (Table 4). The data in Table 4 indicate that the pyridyl nitrogen in the 3-position is important, and that additional nitrogens in the ring are generally detrimental to activity. The potent activity of quinoline **26** foreshadows the ability for the pyridine ring to accommodate substituents at the 5- and 6-positions. Interestingly, the 3-methyl-5-isoxazole moiety, which had served well as a replacement for pyridine in ABT-418 (**3**), is a poor substitute for pyridine in A-84543.

Nan-Horng (Stan) Lin and his associates next embarked on exploration of analogues of **11** that incorporate substituents on the pyridine ring.[16] The starting plan was to examine representative halo and alkyl substituents at each open position. Several of the requisite 3-hydroxypyridines (i.e. 2-methyl, 6-methyl, 2-bromo, 5-chloro) were commercially available, and several other target modifications were accessible using the chemistry shown in Figure 6. Thus, syntheses of 4-halo and 4-methyl compounds were accomplished via *ortho* lithiation of protected 3-hydroxypyridine (Figure 6), whereas 6-chloro-3-hydroxy pyridine (or, more properly, 2-chloro-5-hydroxypyridine) was prepared from 2-chloro-5-aminopyridine using known[17] (albeit difficult to reproduce) chemistry (Figure 6). Based on the results in Table 5, the tolerance for substituents on the pyridine ring is in the order of position 5 > 6 > 4 > 2.

MARK W. HOLLADAY and MICHAEL W. DECKER

Table 5. Effects of Substitution of the Pyridyl Ring of A-84543 (**11**) with Chloro or Methyl Substituents

Compound	X	K_i (nM)[a]
11	H	0.15
27	2-Me	22
28	2-Br	451
29	4-Me	4
30	4-Br	28
31	5-Me	0.48
32	5-Cl	0.6
33	5-Br	0.27
34	6-Me	1.3
35	6-Cl	0.6
36	6-Br	1.8

Note: [a]Displacement of [^{3}H]cytisine from rat brain membranes; data from ref. 16.

V. IDENTIFICATION OF ABT-089

Based on the data in Table 3, it can be inferred that des-methyl, R-desmethyl, azetidine, and des-methyl azetidine analogues of the most active analogues in Tables 4 and 5 also should be high-affinity nAChR ligands, and indeed, most of these compounds were eventually prepared and found to have nAChR binding affinity in the expected range (vide infra). Whereas the pyridyl ethers clearly represented an attractive lead series, it is important to point out that other series existed at the same time which competed with the pyridyl ethers for chemistry resources. In particular, two additional interesting series of compounds represented by **37** (K_i = 0.38 nM)[18] and **38** (K_i = 2.7 nM)[19] (Figure 7) also had been discovered at Abbott and were under active investigation. Moreover, during this period, the chemistry team consisted of only six chemists.

Following characterization of a substantial number of compounds from the various series (i.e. pyridyl ethers, the two series represented by **37** and **38**, as well as some additional isoxazole compounds[18,20]), compound **39** (ABT-089), the desmethyl analogue of compound **27** (Table

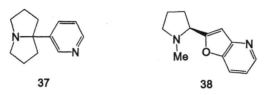

Figure 7. Prototype members of the 8-pyrrolizidine (**37**) and furo[3,2-*b*]pyridine (**38**) series of nAChR ligands discovered at Abbott.

5), emerged as the backup compound to ABT-418.[21–23] Compound **39** was nearly passed over as a candidate for advanced evaluation because of its comparatively low binding affinity (K_i = 18 nM), which is about fivefold weaker than that of ABT-418 and >100-fold weaker than **11**. However, in addition to possessing activity comparable to that of nicotine in rodent assays for cognition enhancement,[21] compound **39** had the important virtue of showing minimal activation of ganglionic-like nAChRs in the IMR-32 human neuroblastoma cell line.[22] Measurement of functional activity in IMR-32 cells had been established as part of the compound screening strategy and was used as an indicator of likely side effects arising from activation of peripheral nAChRs in autonomic ganglia (e.g. effects on the gastrointestinal and cardiovascular systems), and subsequent studies affirmed that ABT-089 was substantially less potent than nicotine or ABT-418 to elicit pressor responses in anesthetized dogs.[23] Moreover, ABT-089 showed good oral bioavailability (ca. 30–70% across rat, dog, and monkey),[23] a property lacking in both nicotine and ABT-418. During the course of profiling ABT-089, a functional assay believed to reflect activation of the putative α4β2 nAChR subtype labeled by [3H]cytisine[24] and [3H]nicotine in rodent brain[25] was established in the project. It was surprising to learn that ABT-089 is only a partial agonist in this assay,[26] and gave some pause to the working assumption that this major brain nAChR subtype was likely to be principally involved in the behavioral effects of ABT-089.

39 (ABT-089)   **40** (-)Epibatidine

VI. FROM ALZHEIMER'S DISEASE TO ANALGESIA: OPPORTUNITY KNOCKS

In the mid-1970s, Daly and coworkers reported isolation of a new substance from the skin of an Ecuadoran frog *Epipedobates tricolor* that showed potency several hundred fold greater than morphine in a screen for analgesic activity. The structure epibatidine (**40**) was first elucidated in the early 1990s, once improved analytical techniques became available,[27] and found to have a number of structural features in common with nicotine. Shortly thereafter, it was reported that the analgesic actions of epibatidine involved activation of neuronal nAChRs.[28,29] Epibatidine is a highly toxic substance, causing death in mice at only about six times the effective dose for analgesia,[30] and eliciting profound cardiovascular effects in dogs at low doses (<5 nmol/kg iv).[31] Although it was far from certain that a compound could be identified that would achieve a useful separation between analgesic and toxic effects, the potential to develop a novel, nonopioid, non-NSAID treatment for pain suggested that an important opportunity had presented itself for the program already focused on nAChR modulators to take a dramatic new direction.

VII. PAIN: CURRENT THERAPIES AND MEDICAL NEED[32-34]

Pain is the most common reason for physician visits. During the course of a year more than 100 million people in the U.S. alone will experience conditions associated with moderate to severe pain. Pain can serve a useful purpose by alerting sufferers to tissue damage and potentially dangerous underlying conditions, but unrelieved pain can disrupt normal function. In addition, millions suffer from neuropathic pain, a condition in which pain perception is distorted by nervous system damage, resulting in pain sensations that are more intense than is appropriate for the stimulus. The extreme condition is one in which chronic pain is experienced even in the absence of noxious stimulation.

Most common agents for treating pain are either nonsteroidal anti-inflammatory drugs (NSAIDs) or opioids. Origins of the use of these compounds date back to the use of willow bark and opium poppies for pain relief. Because there are major deficiencies associated with these types of compounds, the search for new analgesics has largely focused on making incremental improvements within these two compound classes. For example, major drug discovery efforts have made an attempt to limit the gastric irritation produced by NSAIDs or the tolerance and physical dependence liabilities of opioids. A number of anticonvulsant and antidepressant medications are used clinically as well, particularly in the treatment of neuropathic pain, but use of these compounds as analgesics is somewhat limited. Based on the limitations of currently available analgesics and the huge market, there has been an increase in efforts to identify novel classes of analgesics, with a variety of approaches being examined, including neurokinin-1 receptor antagonists, sodium channel blockers, NMDA receptor antagonists, muscarinic cholinergic agonists, and compounds interacting with purinergic neurotransmission.

VIII. IDENTIFICATION OF AZETIDINE PYRIDYL ETHERS WITH ANALGESIC ACTIVITY

During the time when it became apparent that pursuit of novel nAChR modulators for the treatment of pain represented a new and potentially attractive opportunity, the behavioral pharmacology group was mainly occupied with characterizing ABT-089 in advanced behavioral models of cognition enhancement, while the chemistry group was principally engaged in scaling up the synthesis of ABT-089 and other related activities (preparing putative metabolites, supporting synthesis of radiolabeled ABT-089, etc.). Therefore, the early efforts to capitalize on the analgesia opportunity were necessarily limited, and consisted mainly of mechanistic studies of the in vitro and behavioral effects of epibatidine.[30,31,35] Meanwhile, a panel of nAChR ligands representing several different structural series was sent to an outside contract firm for analgesia screening in mouse hot plate paradigms. Among the compounds submitted were several variants in the pyridyl ether series, namely compounds **41–43**. The results indicated that compound **41** showed a significant effect in these screens at 6.2 μmol/kg, whereas compounds **42**, **43**, and members of several other compound series showed no effect at doses up to 62 μmol/kg. Interestingly, the results with **41** later failed to replicate under analgesia screening conditions

41   **42**   **43**

established in-house, but already an impression had been formed that, although analgesic activity was not to be found in every pyridyl ether and certainly not in every potent nAChR ligand, the pyridyl ether series might hold some promise for analgesic activity and should be explored more fully.

Further chemistry effort on the pyridyl ethers was organized according to a spreadsheet exemplified by Table 6, representing a grid of compounds that could be anticipated to exhibit potent nAChR binding affinity based on the data in Tables 3–5. As shown in Table 6, the majority of these analogues do indeed exhibit nAChR affinity in the subnanomolar to low nanomolar range. It should be stressed that it was recognized that binding to the [^{3}H]cytisine site, i.e. the putative $\alpha4\beta2$ nAChR, could not necessarily be expected to be a good predictor of analgesic activity. First, the binding assay provides no information on the effect of ligands on functional responses at the receptor level. Moreover, the possibility that other nAChR subtypes might be involved, including nAChRs in the spinal cord,[35] was well appreciated. Therefore, the [^{3}H]cytisine binding assay mainly served as an approximate guide to the potency of interaction of new ligands with neuronal nAChRs in general. Functional assays for the various nAChR subtypes, which might have been more revealing, mainly utilized native tissue and were often at least as tedious to perform as direct screening for analgesic activity in vivo. For this reason, together with the uncertainty about which subtype should be regarded as the desired site of action, [^{3}H]cytisine binding was generally followed directly by behavioral evaluation in mice, with assessment in in vitro functional assays occurring subsequently for compounds of interest. Table 7 shows activities in the mouse hot-plate assay for the compounds of Table 6. It is apparent that most of the compounds with potent analgesic activity, as revealed by this assay, are found clustered in the lower right portion of the table (cf. positions **F-5**, **G-5**, **I-5**), i.e. are azetidine analogues having a secondary nitrogen atom in the azetidine ring. But not all secondary azetidines with high binding affinity exhibit potent analgesic activity. For example, the 5-chloro substituted compound **D-5** is inactive at doses up to 10-fold higher than the minimally effective dose for **G-5**. Moreover, **D-5** is a full agonist (ca. 120% of nicotine's maximal response) in functional assays at two neuronal nAChR subtypes, the human $\alpha4\beta2$- and human $\alpha3$-containing ganglionic like nAChRs.[36]

Table 6. [³H]Cytisine Binding Affinities (nM) for Pyridyl Ether Compounds of General Structure (Azacycle-CH$_2$O-Het)[a]

		A	B	C	D	E	F	G	H	I
	Azacycle\Het	H	2-Me-3-pyridyl	5-Me-3-pyridyl	5-Cl-3-pyridyl	5-Br-3-pyridyl	6-Me-3-pyridyl	6-Cl-3-pyridyl	6-Br-3-pyridyl	3-F-Ph
1	(S)-NMe-2-pyrrolidinyl	0.15	22	0.48	0.6	0.27	1.3	0.6	1.8	4.9
2	(S)-2-pyrrolidinyl	0.16	18	0.15	0.13	0.85	0.23	0.09		14
3	(R)-2-pyrrolidinyl	0.14	82	0.15	0.25		0.5	0.45		13
4	(S)-NMe-2-azetidinyl	0.45	87	0.96	1.9	2.2	1.7	1.6	2.1	14
5	(S)-2-azetidinyl	0.05	1.4	0.047	0.042	0.18	0.057	0.04	0.021	3.1

Note: [a]Positions on the pyridine ring are numbered such that the azacycle-CH$_2$O-substituent is always at the 3-position.

101

Table 7. Activity in Mouse Hot Plate Assay (Minimum Effective Dose, μmol/kg) for Pyridyl Ether Compounds of General Structure (Azacycle-CH$_2$O-Het)[a]

Azacycle\Het	A	B	C	D	E	F	G	H	I
	H	2-Me-3-pyridyl	5-Me-3-pyridyl	5-Cl-3-pyridyl	5-Br-3-pyridyl	6-Me-3-pyridyl	6-Cl-3-pyridyl	6-Br-3-pyridyl	3-F-Ph
1 (S)-NMe-2-pyrrolidinyl	>6.2	NT	>19	>6.2	>19	62[b]	>6.2	NT	62
2 (S)-2-pyrrolidinyl	>6.2	>62	6.2	>62	>19	>62	>6.2		NT
3 (R)-2-pyrrolidinyl	NT[c]	>62	62	>62	>19	>19	>6.2		62
4 (S)-NMe-2-azetidinyl	>6.2	NT	>62	62[b]	>62	>62	>6.2	>62	>62
5 (S)-2-azetidinyl	>6.2	>62	>62	>6.2	>19	0.62	0.62	>6.2	6.2

Notes: [a]Positions on the pyridine ring are numbered such that the azacycle-CH$_2$O-substituent is always at the 3-position.
[b]Hyperalgesic response.
[c]NT = not tested.

102

The compound at position **G-5** (A-98593) which possesses the α-chloropyridine moiety in common with epibatidine was selected for more detailed evaluation, and was shown to have antinociceptive activity across a number of preclinical pain models. In mice, the compound was active in the hot-plate assay and the abdominal constriction (writhing) assay, suggesting activity against both acute thermal pain and persistent chemical pain.[37] The compound was orally active, and its antinociceptive effects were prevented by mecamylamine, a noncompetitive nAChR antagonist. With twice-daily dosing, A-98593 maintained full efficacy for at least 5 days, whereas reductions in locomotor activity and body temperature observed with acute administration were significantly attenuated under this dosing regimen (A. Bannon and K. Gunther, unpublished data). In rats, the compound was active in the thermal paw withdrawal test and in the formalin test (A. Bannon, P. Curzon, and M. Decker, unpublished data), again displaying activity in acute and persistent pain models. Moreover, A-98593 was active in a spinal nerve ligation model of neuropathic pain (A. Bannon, J. Campbell, and M. Decker, unpublished data). However, A-98593 also showed cardiovascular toxicity in dogs that correlated with its potent in vitro activity in IMR-32 cells, which express nicotinic receptors that resemble those found in autonomic ganglia.[38] A-98593 represented a clear advance relative to epibatidine, but based on its cardiovascular toxicity, it was not regarded as the ideal candidate for clinical development.

IX. (*R*)-AZETIDINES AND IDENTIFICATION OF ABT-594

The knowledge that potent analgesic activity existed in certain azetidine-containing pyridyl ethers did not necessarily elicit a high level of confidence that useful therapeutic agents would be found in this class of compounds. Besides the unacceptable toxicity found in A-98593 (**G-5**), there was a propensity for final compounds in the azetidine series to form by-products during N-deprotection and subsequent manipulations, which caused concerns about the ability to eventually scale up these compounds. Moreover, since analgesic activity had been observed in several analogues in other series, for example compounds related to **37** and **38**, it was a persistent question whether continued work in this series should receive high priority.

Azetidine pyridyl ethers in the (*R*)-stereochemical series had been targets of interest for some time as a natural extension of the structure–activity relationships of Table 3. However, the synthesis of these analogues was hampered by difficulties in reproducing a published route to (R)-azetidinecarboxylic acid (44) from D-methionine (Figure 8).[39] In our hands, the overall yield was low, the cyclization reaction leading to formation of the azetidine ring was capricious, and the final product was partially racemized. Thus, this series shared the potential azetidine stability liabilities described above, and also suffered from further issues of synthetic accessibility. Moreover, pyridyl ether compounds in the (R)-pyrrolidine series had rarely shown behavioral activity comparable to that of corresponding (S)-compounds, and a common presumption was therefore that (R)-azetidines would likely show similarly unimpressive in vivo activity. Nevertheless, the opinion prevailed that the stability and synthetic issues were not insurmountable should the compounds show promising activity, so the efforts of one chemist were maintained on (R)-azetidines.

To satisfy the immediate needs for precursor compound 44 to permit structure–activity studies, the synthetic route from Figure 8 was coupled with an optical resolution of partially racemized 44 (as the N-Cbz derivative) using a published method.[40] The first batch of 44 obtained in this way was sufficient for preparation of prototype analogue 45 (Table 8), which when tested in the mouse hot plate was inactive at doses up to 6.2 μmol/kg ip. A subsequent batch of 44 produced sufficient material for three additional analogues, and the enantiomers (49 and 1) of the compounds that had shown analgesic activity in the (S)-stereochemical series were targeted, together with compound 48. As it turned

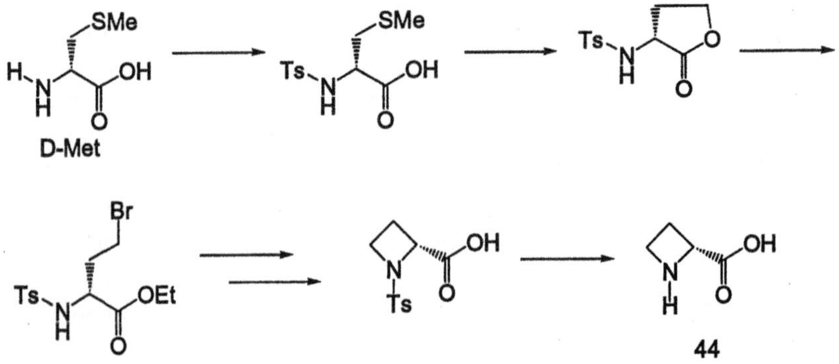

Figure 8. Route to (R)-2-azetidinecarboxylic acid from D-methionine (adapted from ref. 39).

Table 8. Effects of Substitution of the Pyridyl Ring of (*R*)-Azetidine
Pyridyl Ether Analogues[a]

Compound	X	K_i (nM)[b]	MED[c] μmol/kg[d]
45	H	0.05	>6.2
46	2-Cl	85	62
47	5-Me	0.13	62
48	5-Cl	0.12	62
49	6-Me	0.07	6.2
1	6-Cl	0.04	0.62
50	6-Br	0.17	0.62

Notes: [a]Data from ref. 41.
[b]Displacement of [³H]cytisine from rat brain membranes.
[c]MED = minimal effective dose.
[d]Activity in mouse hot plate assay.

out, compound **1** possessed analgesic activity in the hot-plate assay comparable to that of its (*S*)-enantiomer A-98593 (**G-5**, Table 6).[38] Compound **49** also was active, but was somewhat weaker than its (*S*)-enantiomer (**F-5**, Table 6), whereas 5-chloro compound **48** was weaker still. Most important, it was learned shortly thereafter that the activity of compound **1** at human ganglionic-like receptors was attenuated relative to compound **G-5**, suggesting the possibility that compound **1** might show reduced liability for side effects resulting from activation of nAChRs in autonomic ganglia.[38] Further biological profiling, as described subsequently, eventually demonstrated that compound **1**, later code-named ABT-594, possessed properties that made it suitable to be recommended for clinical studies.

X. FURTHER STRUCTURE–ACTIVITY STUDIES IN THE AZETIDINE SERIES

Many additional azetidine analogues were prepared with various substituents on the pyridine ring.[41] In addition to the analogues shown in Tables 6–8, representative further examples are shown in Table 9. Compounds **51–60** include several enantiomeric pairs, and it is apparent that there

Table 9. Effects of Substitution of the Pyridyl Ring of Azetidine Pyridyl Ether Analogues[a]

Compound	Stereochemistry	X	K_i (nM)[b]	MED[c] μmol/kg[d]
51	S	6-OMe	1.3	>62
52	R	6-OMe	0.67	>62
53	S	6-F	0.057	6.2
54	R	6-F	0.066	1.9
55	S	5-Cl, 6-Cl	0.023	0.62
56	R	5-Cl, 6-Cl	0.050[e]	6.2[e]
57	S	5-Br, 6-Cl	0.015[e]	1.9[e]
58	R	5-Br, 6-Cl	0.023	0.62
59	S	2-Cl	2.3	62
60	R	2-Cl	85	62
61	S	6-CN	1.9	62
62	S	6-Et	19	>6.2

Notes: [a]Data from ref. 41 unless otherwise indicated.
[b]Displacement of [^{3}H]cytisine from rat brain membranes.
[c]MED = minimal effective dose.
[d]Activity in mouse hot plate assay.
[e]Unpublished data.

are close parallels in both the binding affinities and analgesic activities between the two stereochemical series. In addition, analogues with substituents other than halo or methyl at the 6-position [e.g. 6-OMe (**51** and **52**), 6-CN (**61**), 6-Et (**62**)] show comparatively weak activity, as do 2-chloro analogues **59** and **60**. The data for 6-ethyl compound **62** suggest that the 6-position is sensitive to substituent size, since compared to the corresponding 6-methyl compound (compound **F-5**, Tables 6 and 7), the 6-ethyl compound possesses much lower binding affinity. 5,6-Disubstituted compounds **55–58** show high binding affinity and potent analgesic activity, which may be compared with the lack of potent analgesic activity for corresponding 5-monosubstituted analogues **D-5** and **E-5** (Tables 6 and 7).

In addition, analogues of A-98593 or ABT-594 containing one (**63**) or two (**64**) methyl substituents on the azetidine ring or an additional

Table 10. Effects of Azetidine Substitution and Linker Homologation on Binding Affinity and Analgesic Activity[a]

Compound	Structure	K_i (nM)[b]	MED[c] μmol/kg[d]
63		7.6	>62
64		37	>62
65		11[e]	62[e]

Notes: [a]Data from reference 41 unless otherwise indicated.
[b]Displacement of [^{3}H]cytisine from rat brain membranes.
[c]MED = minimal effective dose.
[d]Activity in mouse hot plate assay.
[e]Unpublished data.

methylene unit in the interring linking chain (**65**) were prepared (Table 10). These subtle modifications were sufficient to cause profound decreases in biological activity.

XI. PHARMACOLOGICAL PROFILE OF ABT-594

ABT-594 (**1**) is a potent agonist at nAChRs and has antinociceptive activity in several rodent models of acute, persistent, and neuropathic pain. ABT-594 has affinity for the [^{3}H]cytisine binding site comparable to that of epibatidine (K_i = 0.037 nM and 0.043 nM for ABT-594 and (±)-epibatidine, respectively). However, ABT-594 is a less potent inhibitor of α-bungarotoxin binding than epibatidine in rat brain (60 times less potent than epibatidine) and in *Torpedo* electroplax (3000 times less potent than epibatidine), which represent, respectively, the α7-containing neuronal nAChR and the neuromuscular-type nAChR.[42,43] In cell lines expressing α3-containing or α4-containing nAChRs, ABT-594 is, respectively, 29-fold and threefold weaker than epibatidine.[42] Still, ABT-594 appears to display greater overall selectivity than does epibatidine.

As a result of improved selectivity, ABT-594 is safer than epibatidine in in vivo testing. The separation between antinociceptive doses and lethal doses in mice is fivefold greater for ABT-594 than it is for epibatidine.[37] Similarly, ABT-594 has less seizure liability than epibatidine in mice and displays less cardiovascular toxicity in dogs.[37,38]

In mice, ABT-594 produces antinociceptive effects in both the hot-plate test and the abdominal constriction assay. In rats, similar results are obtained in the thermal paw withdrawal test and the formalin test. The efficacy of ABT-594 in these tests is similar to or better than that of morphine, and ABT-594 is at least 30 times more potent than morphine. In addition, ABT-594 is effective in the spinal nerve ligation (Chung) model of neuropathic pain. In all of these models, ABT-594 shows a rapid onset of action and activity after oral administration.

Thus, results from preclinical testing suggest that ABT-594 exhibits the broad spectrum of activity observed with opioids. The analgesic effects of ABT-594, however, do not appear to be mediated through interactions with opioid systems. Effects of ABT-594 in the mouse hot plate cannot be prevented by administration of naltrexone, a opioid receptor antagonist.[37] Similar results have been obtained in rats. Moreover, the in vivo effects of ABT-594 differ from those of morphine in several ways. The antinociceptive effects of ABT-594 in rats is maintained after 5 days of twice-daily dosing, a dosing regimen that produces tolerance to the analgesic effects of morphine.[44] ABT-594 also differs from morphine in that it does not produce hypercapnia and does not produce the overt physical withdrawal effects.[37,43] Furthermore, EEG recordings from rats given ABT-594 do not display a pattern consistent with inattention and sedation that one sees with morphine.[44]

Although a peripheral site of action is possible, the evidence collected to this point suggests that the antinociceptive effects of ABT-594 are mediated by interactions with nAChRs within the central nervous system (CNS). Effects in the mouse hot plate can be prevented by mecamylamine, but not by hexamethonium, a nAChR antagonist that does not readily enter the CNS.[37] Similar results were obtained in rats using the thermal paw withdrawal test, and icv injection of the long-lasting nAChR antagonist chlorisondamine prevents the antinociceptive effects of ABT-594 in this test and in the formalin test.[44] Thus, selective blockade of CNS nAChRs is sufficient to prevent the effects of ABT-594. Moreover, direct injection of ABT-594 into the nucleus raphe magnus of the brainstem has antinociceptive effects.[43,45] This finding and the observations that systemic administration of ABT-594 activates neurons in the NRM and that serotonergic neurons within this region express the $\alpha 4$ subunit of the nAChR[46] suggest that activation of serotonergic projections from the brainstem to the spinal cord may be an important mediator of the effect. Other regions in the CNS might also be important, however, since destruction of serotonergic neurons in the NRM attenuates the antinociceptive effect of ABT-594 but does not completely prevent it.[45] ABT-594, for example, may have effects on pain transmission through effects on nAChRs directly within the spinal cord. The compound can attenuate capsaicin-induced release from spinal cord slices of CGRP and substance P, neuropeptides associated with pain transmission, and can decrease responses of spinal cord neurons to noxious stimulation of the paw.[42,43]

XII. PHARMACOKINETICS AND METABOLISM

Preclinical pharmacokinetic parameters for ABT-594 are summarized in Table 11. ABT-594 shows good oral bioavailability across species, with oral half-lives ranging between 1.4 and 4.2 h, and rapidly enters the brain, exhibiting a brain to plasma ratio of about 2 within 1.5 h (K. Marsh, et al., unpublished data). In vitro metabolism studies using liver homogenates from several species indicated very little metabolism of ABT-594, whereas positive controls were extensively metabolized. In radiotracer studies in vivo, ABT-594 was mainly excreted unchanged in the urine (J. Ferrero, B. Surber, et al., unpublished data).

Table 11. Summary of Preclinical Pharmacokinetic Parameters for ABT-594 (**50**)[a]

Species	$t_{1/2}$ (h), iv	$t_{1/2}$ (h), po	Bioavailability (%)
Mouse	0.4	1.4	78
Rat	1.5	2.0	61
Dog	4.7	4.2	31
Monkey	1.4	ND	80

Note: [a]Unpublished data.

XIII. PROCESS CHEMISTRY DEVELOPMENT FOR ABT-594

The need for a scaleable synthesis of precursor **44**, or a suitable derivative, remained an important issue, both to permit further structure–activity studies, and to enable preparation of larger quantities of ABT-594
for advanced profiling and clinical studies. Thus, a program to address
this problem was initiated within the medicinal chemistry team. It was
recognized that D-aspartic acid contained all of the requisite functionality, and several routes from this starting material were investigated.
The most promising route (Figure 9)[47] proved to be one that proceeded
through key β-lactam intermediate **66**, a compound already known from
a Merck thienamycin synthesis.[48] Product alcohol **67** can be directly
carried forward to ABT-594.[47] The route in Figure 9 has since been
successfully applied to preparations of ABT-594 on multi-kilogram
scale.

XIV. FUTURE PROSPECTS

ABT-594 evolved from several years of research at Abbott Laboratories
on nicotinic acetylcholine receptor modulators, directed initially toward
treatment of Alzheimer's disease, and subsequently toward novel agents
for treatment of pain. ABT-594 represents the first of a new generation
of nAChR modulators to be tested as an analgesic agent in human clinical
trials. The outcome of these studies with respect to both efficacy and

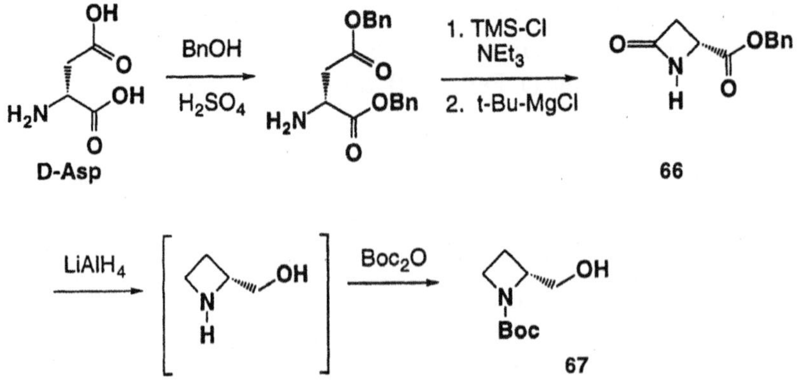

Figure 9. Route to Boc-(*R*)-2-azetidinemethanol from D-aspartic acid
(from ref. 47).

tolerability will help to establish the potential of this mechanism as an alternative to opioids, NSAIDS, and other emerging treatments. If this compound shows promise, it can be expected that additional, improved agents will follow quickly. Moreover, additional efforts to better understand which nAChR subtype(s) may be responsible for nAChR-mediated analgesia surely will ensue.

ACKNOWLEDGMENTS

The work described here represents the efforts of many talented and hard-working colleagues, most of whom are recognized in the reference citations. The decisive support and leadership of Mike Williams and Steve Arneric must be reemphasized, along with the pioneering work of John Daly for the discovery of epibatidine, thereby stimulating our work on nAChRs as potential analgesic agents.

REFERENCES

1. Sargent, P. B. *Ann. Rev. Neurosci.* **1993**, *16*, 403–443.
2. Lindstrom, J.; Anand, R.; Peng, X.; Gerzanich, V.; Wang, F.; Li, Y. *Ann. New York Acad. Sci.* **1995**, *757*, 100–116.
3. Holladay, M. W.; Dart, M. J.; Lynch, J. K. *J. Med. Chem.* **1997**, *40*, 4169–4194.
4. Decker, M. W.; Brioni, J. D.; Bannon, A. W.; Arneric, S. P. *Life Sci.* **1995**, *56*, 545–570.
5. Saunders, J.; MacLeod, A. M.; Merchant, K.; Showell, G.; Snow, R. J.; Street, L. J.; Baker, R. *J. Med. Chem.* **1988**, *31*, 486–491.
6. Orlek, B. S.; Blaney, F. E.; Brown, F.; Clark, M. S. G.; Hadley, M. S.; Hatcher, J.; Riley, G. J.; Rosenberg, H. E.; Wadsworth, H. J.; Wyman, P. *J. Med. Chem.* **1991**, *34*, 2726–2735.
7. Garvey, D. S.; Wasicak, J. T.; Decker, M. W.; Brioni, J. D.; Buckley, M. J.; Sullivan, J. P.; Carrera, G. M.; Holladay, M. W.; Arneric, S. P.; Williams, M. *J. Med. Chem.* **1994**, *37*, 1055–1059.
8. Garvey, D. S.; Wasicak, J. T.; Elliott, R. L.; Lebold, S. A.; Hettinger, A.-M.; Carrera, G. M.; Lin, N.-H.; He, Y.; Holladay, M. W.; Anderson, D. J.; Cadman, E. D.; Raszkiewicz, J. L.; Sullivan, J. P.; Arneric, S. P. *J. Med. Chem.* **1994**, *37*, 4455–4463.
9. Arneric, S.; Anderson, D. J.; Bannon, A. W.; Briggs, C. A.; Buccafusco, J. J.; Brioni, J. D.; Cannon, J. G.; Decker, M. W.; Donnelly-Roberts, D.; Gopalakrishnan, M.; Holladay, M. W.; Kyncl, J.; Marsh, K. C.; Pauly, J.; Radek, R.; Rodrigues, A. D.; Sullivan, J. P. *CNS Drug Rev.* **1995**, *1*, 1–26.
10. Lin, N.-H.; Abreo, M. A.; Gunn, D. E.; Lebold, S. A.; Lee, E. L.; Wasicak, J. T.; Hettinger, A.-M.; Daanen, J. F.; Garvey, D. S.; Campbell, J. E.; Sullivan, J. P.; Williams, M.; Arneric, S. P. *Bioorg. Med. Chem. Lett.* **1999**, *9*, 2747–2752.
11. Abreo, M. A.; Lin, N.-H.; Garvey, D. S.; Gunn, D. E.; Hettinger, A.-M.; Wasicak, J. T.; Pavlik, P. A.; Martin, Y. C.; Donnelly-Roberts, D. L.; Anderson, D. J.;

Sullivan, J. P.; Williams, M.; Arneric, S. P.; Holladay, M. W. *J. Med. Chem.* **1996**, *39*, 817–825.

12. Abood, L. G.; Lu, X.; Banerjee, S. *Biochem. Pharmacol.* **1987**, *36*, 2337–2341.

13. Abood, L. G.; Lerner-Marmarosh, N.; Wang, D.; Saraswati, M. *Med. Chem. Res.* **1993**, *2*, 552–563.

14. Abreo, M. A.; Gunn, D. E.; Lin, N.-H.; Elliott, R. E.; Garvey, D. S.; Lebold, S. A.; Wasicak, J. T. In Abbott Laboratories: WO, 9408992A1 (1994).

15. Elliott, R. E.; Kopecka, H.; Gunn, D. E.; Lin, N.-H.; Garvey, D. S.; Ryther, K. B.; Holladay, M. W.; Anderson, D. J.; Campbell, J. E.; Sullivan, J. P.; Buckley, M. J.; Gunther, K. L.; O'Neill, A. B.; Decker, M. W.; Arneric, S. P. *Bioorg. Med. Chem. Lett.* **1996**, *6*, 2283–2288.

16. Lin, N.-H.; Gunn, D. E.; Li, Y.; He, Y.; Ryther, K. B.; Kuntzweiler, T.; Donnelly-Roberts, D. L.; Anderson, D. J.; Campbell, J. E.; Sullivan, J. P.; Arneric, S. P.; Holladay, M. W. *Bioorg. Med. Chem. Lett.* **1998**, *8*, 249–254.

17. Effenberger, F.; Krebs, A.; Willrett, P. *Chem. Ber.* **1992**, *125*, 1131–1140.

18. Wasicak, J. T.; Garvey, D. S.; Holladay, M. W.; Lin, N.-H.; Ryther, K. B. In Abbott Laboratories: US, 5733912 (1998).

19. Elliott, R. L.; Ryther, K. B.; Anderson, D. J.; Piattoni-Kaplan, M.; Kuntzweiler, T. A.; Donnelly-Roberts, D.; Arneric, S. P.; Holladay, M. W. *Bioorg. Med. Chem. Lett.* **1997**, *7*, 2703–2708.

20. Lebold, S. A.; Hettinger, A.-M.; Garvey, D. S.; Fitzgerald, M. A.; Holladay, M. W.; Pavlik, P.; Martin, Y. C.; Anderson, D. J.; Campbell, J. E.; Donnelly-Roberts, D.; Piattoni-Kaplan, M.; Sullivan, J. P.; Arneric, S. P. *Abst. 210th Amer. Chem. Soc.* **1995**, MEDI-144.

21. Decker, M. W.; Bannon, A. W.; Curzon, P.; Gunther, K. L.; Brioni, J. D.; Holladay, M. W.; Lin, N.-H.; Li, Y.; Daanen, J. F.; Buccafusco, J. F.; Prendergast, M. A.; Jackson, W. J.; Arneric, S. P. *J. Pharmacol. Exp. Ther.* **1997**, *283*, 247–258.

22. Lin, N.-H.; Gunn, D. E.; Ryther, K. B.; Garvey, D. S.; Donnelly-Roberts, D. L.; Decker, M. W.; Brioni, J. D.; Buckley, M. J.; Rodrigues, A. D.; Marsh, K. G.; Anderson, D. J.; Buccafusco, J. J.; Pendergast, M. A.; Sullivan, J. P.; Williams, M.; Arneric, S. P.; Holladay, M. W. *J. Med. Chem.* **1997**, *40*, 385–390.

23. Arneric, S. P.; Campbell, J. E.; Carroll, S.; Daanen, J. F.; Holladay, M. W.; Johnson, P.; Lin, N.-H.; Marsh, K. C.; Peterson, B.; Qui, Y.; Roberts, E. M.; Rodrigues, A. D.; Sullivan, J. P.; Trivedi, J.; Williams, M. *Drug Devel. Res.* **1997**, *41*, 31–43.

24. Pabreza, L. A.; Dhawan, S.; Kellar, K. J. *Mol. Pharm.* **1990**, *39*, 9–12.

25. Marks, M. J.; Farnham, D. A.; Grady, S. R.; Collins, A. C. *J. Pharmacol. Exp. Ther.* **1993**, *264*, 542–552.

26. Sullivan, J. P.; Donnelly-Roberts, D.; Briggs, C. A.; Gopalakrishnan, M.; Hu, I.; Campbell, J. E.; Anderson, D. J.; Piattoni-Kaplan, M.; Molinari, E.; McKenna, D. G.; Gunn, D. E.; Lin, N.-H.; Ryther, K. B.; He, Y.; Holladay, M. W.; Williams, M.; Arneric, S. P. *J. Pharmacol. Exp. Ther.* **1997**, *283*, 235–246.

27. Spande, T. F.; Garaffo, H. M.; Edwards, M. W.; Yeh, H. J. C.; Pannell, L.; Daly, J. W. *J. Am. Chem. Soc.* **1992**, *114*, 3475–3478.

28. Qian, C.; Li, T.; Shen, T. Y.; Libertine-Garahan, L.; Eckman, J.; Biftu, T.; Ip, S. *Eur. J. Pharmacol.* **1993**, *250*, R13–R14.

29. Badio, B.; Daly, J. W. *Mol. Pharm.* **1994**, *45*, 563–569.

30. Sullivan, J. P.; Decker, M. W.; Brioni, J. D.; Donnelly-Roberts, D.; Anderson, D. J.; Bannon, A. W.; Kang, C.-H.; Adams, P.; Piattoni-Kaplan, M.; Buckley, M.; Gopalakrishnan, M.; Williams, M.; Arneric, S. P. *J. Pharmacol. Exp. Ther.* **1994**, *271*, 624–631.

31. Sullivan, J. P.; Briggs, C. A.; Donnelly-Roberts, D.; Brioni, J. D.; Radek, R. J.; McKenna, D. G.; Campbell, J. E.; Arneric, S. P.; Decker, M. W.; Bannon, A. W. *Med. Chem. Res.* **1994**, *4*, 502–516.

32. Dray, A.; Urban, L.; Dickenson, A. *Trends Pharmacol. Sci.* **1994**, *15*, 190–197.

33. Dray, A.; Urban, L. *Annu. Rev. Pharmacol. Toxicol.* **1996**, *36*, 253–280.

34. Barratt, S. M. G. *International Anesthesiology Clinics* **1997**, *110*, 27–47.

35. Sullivan, J. P.; Bannon, A. W. *CNS Drug Rev.* **1996**, *2*, 21–39.

36. Unpublished data. Assays measured $^{86}Rb^+$ flux and were performed as described in ref. 11.

37. Decker, M. W.; Bannon, A. W.; Buckley, M. J.; Kim, D. J. B.; Holladay, M. W.; Ryther, K. B.; Lin, N.-H.; Wasicak, J. T.; Williams, M.; Arneric, S. P. *Eur. J. Pharmacol.* **1998**, *346*, 23–33.

38. Holladay, M. W.; Wasicak, J. T.; Lin, N.-H.; He, Y.; Ryther, K. B.; Bannon, A. W.; Buckley, M. J.; Kim, D. J. B.; Decker, M. W.; Anderson, D. J.; Campbell, J. E.; Kuntzweiler, T. A.; Donnelly-Roberts, D. L.; Piattoni-Kaplan, M.; Briggs, C. A.; Williams, M.; Arneric, S. P. *J. Med. Chem.* **1998**, *41*, 407–412.

39. Miyoshi, M.; Sugano, H.; Fujii, T.; Ishihara, T.; Yoneda, N. A. *Chem. Lett.* **1973**, 5–6.

40. Rodebaugh, R. M.; Cromwell, N. H. *J. Heterocycl. Chem.* **1969**, *6*, 993–994.

41. Holladay, M. W.; Bai, H.; Li, Y.; Lin, N.-H.; Daanen, J. F.; Ryther, K. B.; Wasicak, J. T.; Kincaid, J. F.; He, Y.; Hettinger, A.-M.; Huang, P.; Anderson, D. J.; Bannon, A. W.; Buckley, M. J.; Campbell, J. E.; Donnelly-Roberts, D. L.; Gunther, K. L.; Kim, D. J. B.; Kuntzweiler, T. A.; Sullivan, J. P.; Decker, M. W.; Arneric, S. P. *Bioorg. Med. Chem. Lett.* **1998**, *8*, 2797–2802.

42. Donnelly-Roberts, D. L.; Puttfarcken, P. S.; Kuntzweiler, T. A.; Briggs, C. A.; Anderson, D. J.; Campbell, J. E.; Manelli, A.; Piatonni-Kaplan, M.; McKenna, D. G.; Wasicak, J. T.; Holladay, M. W.; Williams, M.; Arneric, S. P. *J. Pharmacol. Exp. Ther.* **1998**, *285*, 777–786.

43. Bannon, A. W.; Decker, M. W.; Holladay, M. W.; Curzon, P.; Donnelly-Roberts, D.; Puttfarcken, P. S.; Bitner, R. S.; Diaz, A.; Dickenson, A. H.; Porsolt, R. D.; Williams, M.; Arneric, S. P. *Science* **1998**, *279*, 77–81.

44. Bannon, A. W.; Decker, M. W.; Curzon, P.; Buckley, M. J.; Kim, D. J. B.; Radek, R. J.; Lynch, J. K.; Wasicak, J. T.; Arnold, W. H.; Holladay, M. W.; Arneric, S. P. *J. Pharmacol. Exp. Ther.* **1998**, *285*, 787–794.

45. Decker, M. W.; Curzon, P.; Holladay, M. W.; Nikkel, A. L.; Bitner, R. S.; Bannon, A. W.; Donnelly-Roberts, D. L.; Puttfarcken, P. S.; Kuntzweiler, T. A.; Briggs, C. A.; Williams, M.; Arneric, S. P. *J. Physiology (Paris)* **1998**, *92*, 221–224.

46. Bitner, R. S.; Nikkel, A. L.; Curzon, P.; Arneric, S. P.; Bannon, A. W.; Decker, M. W. *J. Neurosci.* **1998**, *18*, 5426–5432.

47. Lynch, J. K.; Holladay, M. W.; Ryther, K. B.; Bai, H.; Hsiao, C.-N.; Morton, H. E.; Dickman, D. A.; Arnold, W.; King, S. A. *Tetrahedron: Asymmetry* **1998**, *9*, 2791–2794.

48. Salzmann, T. N.; Ratcliffe, R. W.; Christensen, B. G.; Bouffard, F. A. *J. Am. Chem. Soc.* **1980**, *102*, 6161–6163.

DISCOVERY AND PRECLINICAL EVALUATION OF NOVEL DOPAMINE PARTIAL AGONISTS AS ANTIPSYCHOTIC AGENTS

David J. Wustrow

Advances in Medicinal Chemistry
Volume 5, pages 115–158.
Copyright © 2000 by JAI Press Inc.
All rights of reproduction in any form reserved.
ISBN: 0-7623-0593-2

ABSTRACT

Antagonism of DA D2 receptors is thought to be the primary molecular mechanism for the efficacy of existing antipsychotic agents. This efficacy is believed to arise from blockade of DA D2 receptors in the mesolimbic and prefrontal brain regions. However blockade of DA D2 receptors in the striatum and caudate is thought to be responsible for the severe side effects associated with antipsychotic therapy including EPS and TD. Neurons that synthesize and release DA have DA autoreceptors on their presynpatic terminals and cell bodies. Stimulation of these autoreceptors by DA results in a negative feedback signal attenuating the synthesis and release of DA from these neurons. It has been hypothesized that partial agonists selective for DA autoreceptors could inhibit DA neurotransmission and therefore have antipsychotic efficacy. This review details the structure–activity relationships of DA partial agonists selective for DA D2 autoreceptors. Compounds having D2 and D3 affinity or D2 and 5-HT1A activity are also discussed. Compounds with the appropriate level of partial agonist activity were shown to have better efficacy/side effect profiles in primate models than standard DA D2 antagonists.

I. INTRODUCTION

A. DA Antagonist Antipsychotic Agents

Attenuation of brain dopamine (DA) neurotransmission has been widely recognized as a useful mechanism for the treatment of the psychotic symptoms of schizophrenia.[1-3] Antagonism of DA receptors is a major component of the mechanism of action of classical antipsychotic agents such as haloperidol and chlorpromazine as well as the newer "atypical" agents such as clozapine,[4,5] risperidone,[6] sertindole,[7] and olanzapine.[8,9] However a significant portion of schizophrenic patients do not respond to DA antagonist therapy and their use is often limited by a variety of severe side effects including extrapyramidal side effects (EPS) and tardive dyskinesia (TD).[10-12] While atypical antipsy-

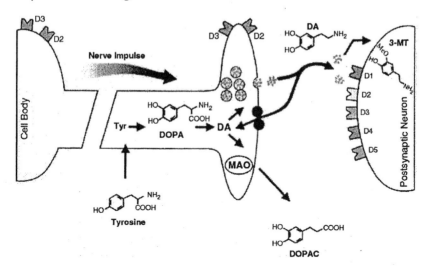

Figure 1. Schematic representation of a DA synapse.

chotics have a somewhat decreased propensity for causing EPS and TD, these side effects are still not uncommon with the newer agents.[13] An exception to this is the atypical antipsychotic agent clozapine that has broad efficacy and causes little or no EPS and TD. However its use is limited because of its potential for causing the potentially fatal blood disorder agranulocytosis.[14] Thus the effective control of schizophrenia in many cases remains an unmet medical need and the psychopharmacological research community still seeks alternatives to the existing DA antagonist medications.[15]

A schematic representation of a brain dopamine synapse is depicted in Figure 1. The five DA receptor subtypes can be divided into two major groups: the D1 family (D1 and D5) which stimulate cyclic AMP formation by increasing the activity of adenylyl cyclase, and the D2 family (D2, D3, and D4) which inhibit adenylyl cylase.[16-18] Existing functional and immunohistochemical evidence suggests all five receptor subtypes are expressed postsynaptically although not necessarily on the same neurons. The efficacy of current antipsychotic agents is thought to occur through blockade of the D2 receptor family in general with the efficacy being best correlated with affinity for the DA D2 receptor subtype.[19] While blockade of DA D2 receptors on postsynaptic nerve terminals in mesolimbic brain regions are believed to be responsible for the efficacy of current antipsychotics, excessive blockade of DA

D2 receptors in striatal regions by these agents may result in EPS and TD.[20]

B. DA Autoreceptor Hypothesis

DA receptors are also found presynaptically on terminals on neurons that synthesize and release the neurotransmitter DA. These presynaptic DA autoreceptors are believed to act as a negative feedback mechanism such that when DA is bound to the receptor the synthesis and release of this neurotransmitter is inhibited.[21–23] DA autoreceptors that regulate the firing rate of DA neurons are located on cell bodies.[24] Several studies indicate presynaptic DA receptors may be both D2 and D3 subtypes.[25,26] However the exact role of the D3 receptor in the CNS is still unclear because truly selective ligands for D3 vs. D2 receptors have been elusive.[27] Because of the lack of selectivity of most DA autoreceptor agonists, a contribution of putative D3 autoreceptors cannot be ruled out. However, the D2 subtype is generally thought to play the major role in DA autoreceptor pharmacology.[28–30]

DA D2 agonists that act selectively at presynaptic DA D2 autoreceptors could decrease the amount of DA synthesized and released into the synapse without completely blocking DA neurotransmission. This would provide a different method to modulate DA neurotransmission without complete postsynaptic receptor blockade. This agonist stimulation must be specific for presynaptic DA D2 autoreceptors as stimulation of postsynaptic DA D2 receptors could exacerbate symptoms of schizophrenia. Postsynaptic and presynaptic DA D2 receptors have identical amino acid sequences.[31,32] At first glance this would make pharmacological differentiation impossible. However, a large body of evidence suggests that DA D2 agonists more potently activate autoreceptors in vivo than postsynaptic receptors.[33] This selectivity of DA D2 agonists and partial agonists for presynaptic receptors reflects the larger receptor reserves on presynaptic receptor fields compared to postsynaptic sites.[34,35] Accordingly, molecules having the appropriate level of partial agonist activity at DA D2 receptors could act as agonists or partial agonists at DA D2 autoreceptors and inhibit synthesis but would not be able to stimulate postsynaptic receptors at therapeutically relevant concentrations. Such an agent would have the desired selectivity for selective autoreceptor activation leading to decreased levels of DA in the synapse. This mechanism would reduce the symptoms of schizophrenia without inducing the side effects associated with the postsynaptic blockade caused by DA antagonists.[36]

Compounds were screened for selective DA D2 autoreceptor agonist activity in a number of ways. Binding studies were carried out on rat whole brain preparations and later on cells transfected with cloned human DA D2 receptors by assessing test compounds' abilities to displace DA D2 radioligands such as [^{3}H]haloperidol and [^{3}H]spiperone.[37,38] Once affinities for DA D2 receptors were established, the behavioral and neurochemical effects of compounds were studied for evidence of the inhibition of DA synthesis. Blockade of inhibitory inputs to DA neurons with gamma-butyrolactone (GBL) caused an increase in DA synthesis as assessed by measuring increases in DOPA accumulation after decarboxylase inhibition with NSD 1015 (Figure 1)[39,40] DA neuronal firing rates were assessed in anesthetized rats and decreases observed after drug administration were indicative of DA autoreceptor stimulation.[41] In mice and rats, inhibition of spontaneous locomotor activity can be measured after blockade of DA neurotransmission either by DA antagonism or selective stimulation of DA D2 autoreceptors.[42–44] Compounds that activate both pre- and postsynaptic receptors tend to cause stimulation of locomotor activity. Compounds that appeared to selectively inhibit DA D2 autoreceptors were evaluated for their ability to inhibit the Sidman avoidance responding in squirrel monkeys. This is thought to be an excellent predictor of both antipsychotic efficacy and potency in humans.[45,46] Throughout the project compounds were also evaluated for their propensity for causing EPS in monkeys sensitized with haloperidol.[46,47]

II. PROTOTYPICAL DA AGONISTS

A. Classical DA Agonists

Initial attempts to design DA autoreceptor agonists began with molecules having structural elements that directly mimicked DA. Examples in Figure 2 include apomorphine (1),[48] talipexole (2),[49,50] quinpirole (3),[51] preclamol (4),[52–54] U-68553 (5)[55,56] and PD 128483 (6).[57] Early studies with dopamine agonists such as 1 and 2 showed these molecules not to be efficacious as antipsychotic agents and, in fact, in some cases schizophrenic symptoms were exacerbated.[58] This exacerbation of schizophrenic symptoms was most likely due to the high level of intrinsic agonist activity for the DA D2 receptor these compounds possessed.

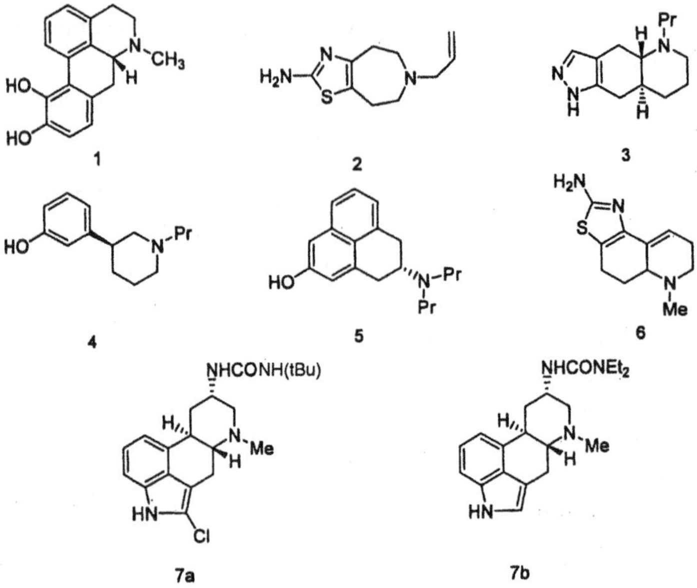

Figure 2. Classical DA agonists.

Because of their high level of intrinsic activity they stimulated postsynaptic receptors as well as the more sensitive presynaptic autoreceptors. On the other hand, compounds with very low intrinsic activity such as SDZ 208-912 (**7a**) and terguride (**7b**)[59] have biological activity which resembles dopamine antagonists and produce EPS in the clinic (Figure 2).[58] All of these compounds have structural elements which directly mimic DA in that they contain an amino group tethered by a two carbon spacer to an aryl or heteroaryl ring having functionality which can form a hydrogen bond with the DA D2 receptor. Models of the interactions of some of these classical agonists with the DA D2 receptor have been proposed in which an overlay with DA is assumed.[60,61]

B. Early Nonclassical DA Agonists

By the middle of the 1980s a second class of DA D2 agonists began to emerge. Although compounds in this class contained aryl and amino functionalities, their exact overlap with dopamine was less clear. An early compound of this type was indole-tetrahydropyridine roxindole (**8**; Figure 3).[62] It has been suggested that the indole ring of **8** might mimic the catechol ring of DA. However, the flexibility of the four-carbon tether

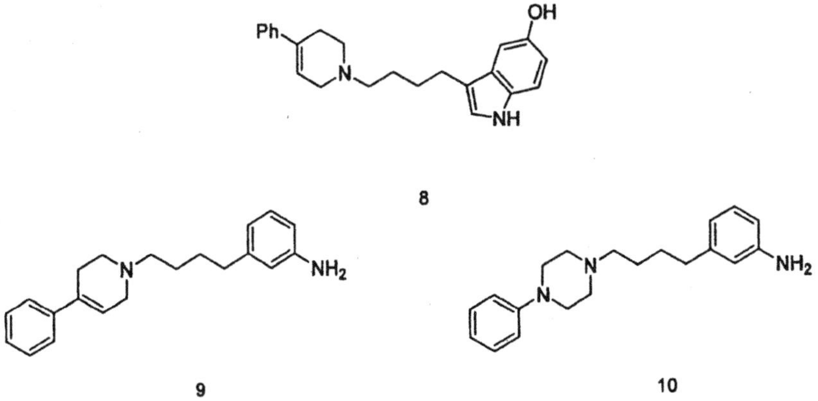

Figure 3. Early nonclassical DA agonists.

between the indole and amino function makes it difficult to determine with certainty the distance between the amine and indole function in the active conformation.

Early efforts from our laboratories concentrated on creating molecules that combined aryl functionality capable of forming hydrogen bonds with aryl piperazine or aryl tetrahydropyridine groups. Representative of this group are the anilines **9** and **10** (Figure 3) which had affinities for DA D2 receptors as measured by their ability to displace [³H]haloperidol from rat brain tissue.[63] Autoreceptor agonist activity was evidenced by the inhibition of DA neuronal firing in vivo and DA synthesis in the rat striatum (Table 1). Stimulation of locomotor activity in 6-OHDA lesioned rats is one of the most sensitive models of postsynaptic DA activity.[64,65] In this model DA agonists such as apomorphine and roxin-

Table 1. Aniline DA Agonists

Pharmacological Assay	9	10
[³H]Haloperidol Receptor Binding (IC$_{50}$ nM)	67	138
Percent Inhibition of DA Neuronal Firing (2.5 mg/kg ip)	89	85
Inhibition of DA Synthesis in Rat Striatum (ED$_{50}$ mg/kg ip)	7.0	12.0
Inhibition of Locomotor Activity (ED$_{50}$ mg/kg po)	12.1	10.3
Inhibition of Locomotor Activity (ED$_{50}$ mg/kg sc)	0.35	1.8
Reversal of 6-OHDA Induced Depression (ED$_{50}$ mg/kg sc)	0.13	>30
Inhibition of Squirrel Monkey Sidman Avoidance (ED$_{50}$ mg/kg po)	NT[a]	8.5

Note: [a]NT = not tested.

dole cause stimulation of locomotor activity. Compound **9** was active in this model indicative of postsynaptic dopamine receptor activation. However, compound **10** was inactive, suggesting a lower level of intrinsic activity insufficient to stimulate postsynaptic DA D2 receptors. This early study revealed that small structural changes could adjust the DA agonist/partial agonist character of particular members of a series of compounds. Further, this study showed it would be possible to design compounds with the appropriate level of intrinsic activity to show selectivity for DA autoreceptors over postsynaptic receptors.

Compound **10** was assessed in the Sidman avoidance primate model of antipsychotic efficacy and shown to have oral activity although its potency was somewhat less than standards, such as thioridazine (ED_{50} 3.9 mg/kg po). The ability of compounds to cause EPS in primates sensitized to haloperidol appears to correlate with the propensity for causing the motor side effects associated with antipsychotic agents in humans. Unlike the standard antipsychotic agent thioridazine, compound **10** at 5 times the Sidman ED_{50} dose did not induce EPS in squirrel monkeys sensitized to haloperidol. Unfortunately it was also found that at high doses the aniline compounds produced convulsions in animals and therefore could not be developed. But with this early proof of concept in hand, we set out to find other agents with improved oral potency in the primate Sidman model that had little or no propensity for causing EPS in haloperidol-sensitized primates.

In addition to DA agonists and antagonists, a number of aryl piperazines have been described in the literature with adrenergic, noradrenergic, and serotonergic activity. A common explanation of the affinity of this class of receptors is the arylpiperazine moiety is a bioisosteric replacement for the arylethyl amine portion of these neurotransmitters. However, it was postulated[66] that the aryl group attached directly to the piperazine might bind in a position allosteric to the binding site for the neurotransmitters themselves. One reason for this suggestion is that, in addition to the aryl ring attached directly to the piperazine ring, the nature of the pendant aromatic ring plays an important role in determining selectivity and affinity for the various neurotransmitter receptors. Therefore the effect of modifying these aryl regions of the pharmacophore was systematically evaluated for its contribution to DA agonist activity. The benzopyranone ring system was found to be an acceptable replacement for the anilino functionality of the previous series, and a variety of 5-, 6-, and 7-aminoalkoxy-benzopyran-4-ones were synthesized and evaluated for their activity at the DA D2 receptor.[66] A sample of these compounds

Table 2. Benzopyranones

No.	Ar	[^{3}H]Haloperidol Binding: % Inhibition at 100 nM	Inhibition of Mouse Locomotor Activity ED_{50} mg/kg ip	Inhibition of Rat Locomotor Activity ED_{50} mg/kg po	% Reversal of DOPA Accumulation	% Inhibition of DA Neuronal Firing @ 10
11	Phenyl	20	2.4	4.4	56	68
12	2-Chlorophenyl	59	>30	NT[a]	NT	NT
13	3-Chlorophenyl	35	7.7	8.5	(–26)	NT
14	4-Chlorophenyl	10	11.1	18.6	IA	NT
15	4-Fluorophenyl	24	7.0	7.1	(–42)	NT
16	2-Methoxyphenyl	48	7.7	>30	NT	NT
17	4-Methoxyphenyl	25	30	>30	NT	NT
18	2-Tolyl	29	2.0	4.8	IA[b]	NT
19	3-Tolyl	38	5.4	22.3	IA	NT
20	4-Tolyl	36	6.1	13.2	IA	NT
22	2-Pyridyl	IC_{50} = 1000 nM	1.3	1.7	100	93
23	2-Pyrimidyl	0	12.4	15.6	NT	NT
24	2-Pyridazinyl	17	26.3	NT	NT	NT

Notes: [a]NT = not tested.
[b]IA = inactive.

is listed in Table 2. Once again, the DA D2 receptor affinity was evidenced by the compound's ability to displace [^{3}H]haloperidol from rat brain tissue. Inhibition of locomotor activity without stimulation of activity at higher doses was used as a preliminary indicator of selective DA autoreceptor agonist activity or postsynaptic DA antagonist activity. An important part of the paradigm was to show that decreases in locomotor activity were not simply due to ataxia. It has been shown that known antipsychotic agents do not cause motor impairment, as measured by the ability of the animals to cling to an inverted screen, at doses which inhibit locomotor activity.[67] A number of compounds in this series containing a substituted phenyl or heteroaryl piperazine moiety were found to inhibit locomotor activity. The position and nature of the substituent on the phenyl ring did not greatly effect the binding affinity of compounds 11–20. Larger changes in behavioral activity were observed with the unsubstituted analogue, 3-chloro, 4-fluoro, and 2-methyl phenyl analogues (11, 13, 15, and 18) having the greatest activity. A small series of six-membered ring heterocycles were also studied (22–24). Of these heterocycles only the 2-pyridyl analogue 22 had good binding and activity in the behavioral screens.

Compounds from this series with significant in vivo potency were examined for their autoreceptor-like effects on DA synthesis. Their ability to reverse GBL induced increases in DA synthesis was assessed by measuring DOPA accumulation after decarboxylase inhibition with NSD 1015.[39,40] The unsubstituted phenyl piperazine 11 inhibited GBL-induced increases in DOPA levels, suggesting autoreceptor agonist activity while the 3-chloro and 4-fluoro analogues 13 and 15 both caused increased DOPA levels indicative of antagonist activity at the DA D2 receptor. These results suggested that substitution of the phenyl ring of 11 with 4-fluoro or 3-chloro changed a DA partial agonist into an antagonist. The 2-pyridyl analogue 22 was the only compound to cause complete reversal of GBL induced increases in DOPA accumulation.

The two compounds 11 and 22 which caused inhibition of DOPA accumulation were studied for their ability to decrease firing rates of dopaminergic neurons, another hallmark of autoreceptor agonist activity. Both compounds decreased neuronal firing although 22 caused a larger decrease. The electrophysiological result was consistent with the neurochemical result in characterizing both 11 and 22 as DA autoreceptor agonists. The selectivity of the compounds for pre- versus postsynaptic DA D2 receptors was assessed by measurement of stereotypic behavior after coadministering the drugs with the D1 agonist SKF 38393. Postsynaptic DA D2 agonists potently induce repetitive rearing, head-swaying, sniffing, licking, and gnawing when administered with SKF 38393.[68–71] No such stereotyped behavior was observed even at doses 10 times that which inhibited locomotor activity. Compound 22 was also active in the monkey Sidman avoidance test predictive of antipsychotic efficacy. The compound inhibited the Sidman avoidance response at 6 mg/kg po. No signs of EPS were observed in haloperidol-sensitized cebus monkeys at doses over 12-fold higher than the ED_{50}, indicating that the compound would have a very low propensity for causing extrapyramidal side effects in the clinic. Compound 22 was taken into early stage development, but these efforts were discontinued because of toxicology in monkeys.

III. CYCLOALKENYL DA AGONISTS

A. Phenyl Cyclohexenyl Analogues

Up to this point several selective DA autoreceptor agonists had been identified which were comprised of an aromatic ring separated from a 4-phenyl-1,2,3,6-tetrahydropyridine or 4-arylpiperazine by flexible four atom spacers (Figure 1). However, it was unclear which aryl ring was imitating the phenyl ring of dopamine. The flexibility of the linker between the two portions of the molecules contributed to this confusion. To better understand the nature of the active conformation, molecules were prepared that had a more rigid cyclohexenyl spacer and 0–2 carbon atoms between the 2-pyridylpiperazine and aryl functionality (Table 3).[72]

In terms of DA D2 receptor binding and in vivo activity, the compounds having 0- or 2-carbon spacers generally were the most potent. The 0- and 2-carbon analogues also were more potent at inhibiting mouse locomotor activity after ip administration. In comparing the phenyl, pyridyl, and thienyl analogues in both the 0- and 2-carbon series it was clear that the pyridyl analogues **26** and **31** were most potent in vivo. In

Table 3. Arylcyclohexenes

$$Ar—\langle\rangle—(CH_2)_n—N\langle\rangle N—\langle\rangle^{N=}$$

No.	Ar	n	[³H]Spiperone Binding IC$_{50}$, nM	Inhibition of Mouse Locomotor Activity ED$_{50}$ mg/kg ip	Inhibition of Rat Locomotor Activity ED$_{50}$ mg/kg po	% Reversal of Rat Brain DA Synthesis at 10 mg/kg ip
25	Phenyl	0	600	1.0	>30	NT
26	2-Pyridyl	0	1180	0.9	5.8	75
27	2-Thienyl	0	352	9.3	>30	NT
28	Phenyl	1	ca. 10,000	ca. 30	NT[a]	NT
29	2-Thienyl	1	2430	>30	NT	NT
30	Phenyl	2	412	2.1	6.7	50
31	2-Pyridyl	2	128	0.4	3.0	89
32	2-Thienyl	2	436	8.4	13.0	46

Note: [a]NT = not tested.

the 2-carbon series the phenyl analogue **30** also had good activity. All had affinity for DA D2 receptors, inhibited locomotor activity after oral administration in the rat, and decreased brain DA synthesis at low doses. Compound **31**, however, stimulated locomotor activity at doses above 10 mg/kg in rat, suggesting an agonist type interaction with postsynaptic DA D2 receptors. Difficulties were encountered in resolving **26** on a large scale; however, the enantiomers of compound **30** could be resolved and were studied in detail.

As shown in Table 4 both enantiomers bound to DA D2 receptors selectively over D1 receptors as measured using the D2 ligand [^{3}H]N-propyl apomorphine and the D1 radioligand [^{3}H]SCH 23390, respectively. Interestingly both enantiomers also inhibited locomotor activity and decreased DA synthesis and inhibited firing of DA neurons in rats with similar activity and potency. However the (S–)-isomer **33** did not cause stereotypy when administered at high doses with the D1 agonist SCH 38393. In contrast the (R+)-isomer **34** was active in this sensitive measure of postsynaptic D2 receptor stimulation. This study suggested that differences in absolute stereochemistry while not necessarily leading to obvious changes in receptor affinity could lead to differences in intrinsic activity. These findings presaged a more concrete demonstration of the effect of stereochemistry on the intrinsic activity of DA D2 partial agonists.

Having demonstrated that the spacer between the remote aryl group and the 4-phenyl-piperazine could be incorporated into a cyclohexenyl structure, we undertook a more systematic study with a related series of 4-phenyl-1,2,3,6-tetrahydropyridines (**35–40**).[73] Various positions for the attachment of the (phenyltetrahydropyridinyl)methyl group around

Table 4. Enantiomers of **30**

No.	Stereochemistry	[^{3}H]NPA[a] Binding IC_{50}, nM	Inhibition of Rat Locomotor Activity ED_{50} mg/kg po	% Reversal of Rat Brain DA Synthesis at 10 mg/kg ip	% Inhibition of DA Neuronal Firing in Rat at 2.5 mg/kg	% Rats with Stereotypy (dose, mg/kg po)
30	(+/–)	88	6.7	50	41	0
33	S (–) 30	82	6.1	49	38	0
34	R (+) 30	64	9.1	57	58	60

Note: [a]NPA = N-propyl apomorphine.

the cyclohexenyl ring were investigated. As shown in Table 5, the greatest activity was observed when this functionality was attached to the 5-position of cyclohexene ring as in compound **36**. Compound **38** having the methylene group attached in the 3-position, despite being somewhat symmetrical, had weaker activity in both the in vitro binding assay and the in vivo behavioral and neurochemical tests. Further study of 5-substituted cyclohexenyl analogues revealed an interesting contrast with the 4-substituted cyclohexenyl analogues previously described (Table 3). In 4-substituted cyclohexenyl series compound **25** containing the piperazinyl function directly attached to the cyclohexenyl ring and compound **30** with the piperazinyl group attached via a 2-carbon tether were more active than compound **28** having a 1-carbon spacer between the cyclohexenyl and the piperazine ring. The situation is reversed in the 5-substituted cyclohexenyl series. As shown in Table 5, compounds **39** and **40** having a 0- or 2-carbon spacer between the cyclohexenyl and tetrahydropyridine groups were much less active than **36** having the methylene spacer. Based on these observations it is likely that the distance between the phenyl on the cyclohexenyl ring and the basic nitrogen functionality is critical for receptor recognition.

Table 5. Cyclohexenyl Phenyltetrahydropyridines

Compound	Point of Attachment	n	[^{3}H]Spiperone Binding IC_{50}, nM	Inhibition of Mouse Locomotor Activity ED_{50} mg/kg ip	% Reversal of Striatial DOPA Synthesis After GBL
35	6	1	3936	>30	NT[a]
36	5	1	112	1.1	86
37	4	1	1000	12.1	32
38	3	1	420	2.5	34
39	5	0	9207	ca. 30	NT
40	5	2	635	10.9	49

Note: [a]NT = not tested.

A number of other aryl-cyclic amine functionalities were examined as replacements for the 1,2,3,6-tetrahydro-4-phenylpiperidine of **36** (Table 6). Thiophene analogue **41** was found to have somewhat weaker DA D2 receptor affinity. In the 1,4-cyclohexenyl series 2-pyridylpiperazine attached to the phenyl cyclohexene produced potent compounds. The 1,5-cyclohexenyl analogue **42** having a 2-pyridylpiperazine group was also an interesting compound. However, the parent tetrahydropyridine **36** caused greater decreases in DA synthesis than piperazine **42**. Replacement of the 2-pyridyl group with phenyl (**43**) or 2-pyrimidyl (**44**) resulted in compounds with decreased receptor affinity and in vivo activity.

Having determined that compound **36** had an interesting profile, efforts were made to separate the enantiomers of this compound. As seen in Table 7, both enantiomers had similar affinity for the DA D2 receptor. The (R)-enantiomer **45** like the racemate decreased DA synthesis (as measured by DOPA accumulation) consistent with autoreceptor activation. In contrast (S)-enantiomer **46** caused increases in DA synthesis consistent with DA D2 antagonist activity. The difference in apparent in

Table 6. Arylamine Structure–Activity Relationships

Compound	NR_1R_2	[^3H]Spiperone Binding (IC_{50}, nM)	Inhibition of Mouse Locomotor Activity ED_{50} mg/kg ip	% Reversal of Striatal DOPA Synthesis After GBL
36		112	1.1	86
41		170	1.5	NT^a
42		141	0.8	61
43		424	2.1	7.5
44		354	8.5	NT

Note: ^{a}NT = not tested.

vivo intrinsic activity between enantiomers was much greater in this 1,5-cyclohexenyl series than in the 1,4 series which had been studied earlier. Indeed the DA autoreceptor agonist activity of racemate **36** was due to the (*R*)-isomer **45**.

A series of enantiopure analogues of **45** were prepared where the aryl group on the cyclohexene ring had been substituted in the 4-position (Table 7). Of the substituents studied, methyl, fluoro, and methoxy were best tolerated in the DA D2 binding assay. The methyl and methoxy analogues **47** and **50** had activity in the rat behavioral model but decreased DA synthesis to a lesser extent than unsubstituted parent **45**. Conversely the 4-fluoro analogue **49** was significantly less potent than **45** at inhibiting locomotor activity in the rat behavioral paradigm but decreased DA activity to a similar extent. It is possible that **49** might have an intrinsic activity level sufficient to stimulate postsynaptic receptors decreasing its efficacy in inhibiting locomotion. Alternatively differences in metabolic processing of **45** and **49** may be responsible for this difference in activity profile.

Table 7. DA Activity of the Enantiomers and Aryl Analogs

Compound	Configuration	Ar	[^3H]Spiperone Binding IC_{50}, nM	Inhibition of Rat Locomotor Activity ED_{50} mg/kg po	% Reversal of Striatal DOPA Synthesis After GBL
36	R/S	Ph	112	3.0	86
45	R	Ph	53	2.9	66
46	S	Ph	113	3.8	50
47	R	4-MePh	328	5.8	42
48	R	4-ClPh	4634	19.9	NT
49	R	4-FPh	76	17.5	74
50	R	4-MeOPh	100	5.6	47
51	R	4-CF$_3$Ph	2554	NT[a]	NT

Note: [a]NT = not tested.

B. Characterization of CI-1007

Compound **45** was selected for further evaluation and given the designation CI-1007. Binding studies were carried out on D2 receptors from a variety of different sources as well as other neurotransmitter receptors (Table 8).[74] CI-1007 showed a greater than 17-fold increase in affinity for rat striatal DA D2 receptors labeled with the agonist ligand [^3H]N-propylnorapomorphine compared to those same DA D2 receptors labeled with the antagonist ligand [^3H]spiperone. This difference in binding affinity can be explained by the ternary complex model of G-protein-coupled receptors, first postulated to explain the interactions of beta-adrenergic receptors and agonists.[75] Agonists of G-protein-coupled receptors are thought to form a ternary complex with the receptor in an active or high-affinity conformation which is coupled to a G-protein. This ternary complex is responsible for agonist signal transduction. Antagonists interact with and stabilize the low-affinity or inactive form of the

Table 8. In Vitro Binding Profile of CI-1007

HO_2C CO_2H

Receptor	[^3H]Ligand	Tissue	$K_i(nM)$
Rat DA D2	NPA	Rat striatum	3.0
Rat DA D2	Spiperone	Rat Striatum	53.0
Human DA D2L	Spiperone	LZR-1	9.0
Human DA D2L	Spiperone	CHO-K1	25.5
Human DA D3	Spiperone	CHO-K1	16.6
Human DA D4.2	Spiperone	CHO-K1	90.9
Sigma	3-PPP	Guinea Pig Brain	33.0
5-HT1A	8-OH-DPAT	Rat Hippocampus	100
5-HT-2	Ketanserin	Rat Cortex	377
α1 Adrenergic	Prazosin	Rat Cortex	1618
α2 Adrenergic	MK-912	Rat Cortex	413
Phencyclidine	TCP	Guinea Pig Brain	>1600
DA Transporter	GBR-12935	Rat Striatum	>3300
DA- D-1	SCH 23390	Rat Striatum	>10000

receptor uncoupled from the G-protein signal transduction element. Agonist radioligands stabilize and therefore label the high-affinity or active form of the receptor. In light of this argument it would appear that CI-1007 interacts more potently with the high-affinity or agonist conformation of the DA D2 receptor. CI-1007 has similar affinity for the long form of the human DA D2 receptor (DA D2L) and human DA D3 receptors, but had somewhat weaker affinity for the human DA D4.2 receptor subtype and was not found to interact with the DA D1 receptor subtype. Its affinity for serotonergic and adrenergic receptors was also found to be significantly weaker than the standard ligands. The compound also did not display interactions with PCP receptors or the dopamine transporter at physiologically relevant concentrations. Similar results were obtained from the study of CI-1007 with a panel of over 40 additional neurotransmitter receptors, ion channels, and enzymes.

For the first time in the project it became possible to obtain an in vitro measure of intrinsic efficacy and assess compounds agonist properties under conditions similar to those of the binding experiments. To assess in vitro functional activity at DA D2 receptors, CI-1007 was evaluated in GH4C1 cells transfected with the long form of the human DA D2 receptor. Cyclic AMP levels in these cells were measured after forskolin stimulation. Dopamine agonists such as quinpirole were shown to cause a dose-dependent decrease in these cyclic AMP levels as D2 receptors are negatively coupled to adenylyl cyclase. The maximal decrease in cyclic AMP levels induced by **45** and a variety of other DA agonists was

Table 9. Relative Intrinsic Activity of DA D2 Agonists

Test Compound	% Intrinsic Activity[a]
Quinpirole	100
Apomorphine	101
EMD 38362	72
CI-1007	53
(−)3-PPP	44
Terguride	50
SDZ 208-912	11
Haloperidol	0

Note: [a] Intrinsic Activity compared to the full DA agonist quinpirole inhibition of forskolin-stimulated cAMP accumulation in GH4C1 cells transfected with the DA D2 receptor.

assessed. As shown in Table 9, CI-1007 caused a decrease that was roughly half as large as that of the full DA D2 agonists quinpirole and apomorphine. This intrinsic activity level was lower than talipexole (**2**) and roxindole (**9**), similar to terguride (**7b**) and 3-PPP (**4**), and greater than SDZ 208-912 (**7a**). This placed CI-1007 near the middle of this intrinsic activity spectrum that runs from full agonist to antagonist.

In vivo microdialysis studies were carried out to determine the effects of CI-1007 on extraneuronal dopamine release in rats. These studies measure the amount of DA released into the extraneuronal space. Significant decreases in DA levels were observed after a dose of 20 mg/kg ip in the rat. The decreases were larger in the nucleus accumbens region (48% decrease from control levels) than in the striatal region (20% decrease). The reason for this difference is not clear but may indicate a difference in feedback mechanisms controlling DA release between the two regions. CI-1007 also showed an autoreceptor profile by its ability to block DA neuronal firing and inhibit DA synthesis in normal and GBL-treated animals.[41] A dose of 2.5 mg/kg ip completely blocked the spontaneous firing of A9 dopaminergic neurons in rats.

Having established the relative intrinsic activity of CI-1007 compared to other DA agonists in vitro and its neurochemical and electrophysiological effects consistent with autoreceptor activation, the behavioral effects of the compound were examined.[41] At a dose of 30 mg/kg po (30 times the ED_{50} for spontaneous locomotor inhibition) the compound did not induce stereotypy in the rat either alone or in combination with the D1 agonist SK 38393. Consistent with earlier studies, it appeared that CI-1007 acted selectively as an agonist on presynaptic DA D2 receptors.

The compound was next assessed in the monkey Sidman paradigm of antipsychotic efficacy and for its propensity to cause EPS. Like the known antipsychotic agent haloperidol (ED_{50} = 0.5 mg/kg po), CI-1007 inhibited conditioned avoidance responding in squirrel monkeys (ED_{50} = 0.6 mg/kg po). Dopamine agonists of higher intrinsic activity such as apomorphine or roxindole did not inhibit the Sidman avoidance response but instead caused increased lever pressing. The lack of stimulant effects with CI-1007 suggested that it should not cause the psychomotor stimulation observed clinically with higher intrinsic activity agonists.

CI-1007 has similar potency to the DA antagonist haloperidol and the weak partial agonist SDZ 208-912 (**7a**) in the Sidman avoidance assay (Table 10). However CI-1007 caused EPS less frequently in

Table 10. Efficacy and Side Effect Measures in Primates

Compound	Sidman ED_{50} (mg/kg po)	Haloperidol-Sensitized Squirrel Monkey		Haloperidol-Sensitized Cebus Monkeys	
		Dose Mg/kg	#EPD/#Tested	Dose	#EPD/#Tested
Haloperidol	0.5	1.0	3/5	0.125	0/4
		2.0	5/5	0.25	4/4
		3.0	5/5	0.5	4/4
SDZ 208-912	0.2	0.25	2/4	0.25	2/2
		0.5	3/3	0.5	2/2
		1.0	3/3		
45 (CI-1007)	0.6	0.625	0/3	1.25	0/4
		1.25	2/10	2.5	3/6
		2.5	2/5	5.0	2/6
		5.0	5/10	10.0	3/4
		10.0	3/7		
85	1.0	2.5		10	3/4
101	0.28			0.28	0/4
		1.4	2/4[a]	1.4	3/3
		2.8	4/4[a]		

Notes: [a]Emesis occurred at these doses.

squirrel monkeys sensitized to haloperidol. As can be seen in Table 10, CI-1007 caused EPS in about only half of squirrel monkeys studied when they received more than 10 times the Sidman ED_{50} dose. In contrast the DA D2 antagonist haloperidol and the low intrinsic efficacy agonist SDZ 208-912 (**7a**) caused EPS in all monkeys studied at doses approximately 4 and 2 times higher, respectively, than the ED_{50} in the Sidman avoidance paradigm. From these studies it was suggested that CI-1007 had less propensity for causing extrapyramidal side effects at its efficacy dose than the D2 antagonist haloperidol or SDZ 208-912 (**7a**), a dopamine partial agonist with very low intrinsic activity.

Based upon its in vitro and in vivo profile in rodents consistent with antipsychotic activity, its efficacy in a primate model of predictive of antipsychotic activity, and a decreased propensity for causing motor disturbances in haloperidol-sensitized squirrel monkeys, CI-1007 was chosen for development as an antipsychotic agent. During development it was discovered that CI-1007 underwent extensive metabolism by aromatic ring hydroxylation (Figure 4).[76] Monohydroxy metabolites **52** and **53** and the dihydroxy metabolite **54** were isolated and their structures were confirmed by synthesis. Compound **53** and **54** were the major metabolites and both bound with good affinity to the DA D2 receptor. However, neither metabolite was thought to greatly contribute to the pharmacological activities observed in primates since neither major metabolite showed activity in the primate Sidman avoidance paradigm. Oxidation of CI-1007 by cytochrome P-450 (CYP) isozymes is likely responsible for these metabolites. CYP isozyme variations among individuals led to inconsistent blood levels in early clinical studies ultimately leading to the discontinuation of the development of CI-1007.

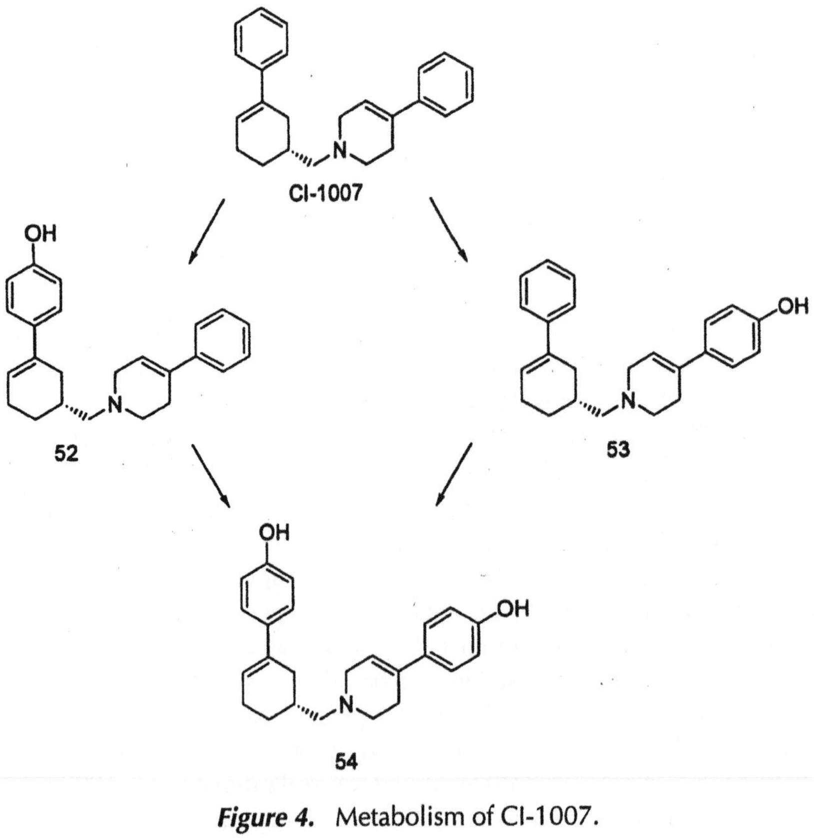

Figure 4. Metabolism of CI-1007.

IV. CYCLOHEXYL BENZAMIDES

A. Initial SAR

Concurrently with the studies in the aryl–cyclohexenyl systems, we undertook the study of amide functionality in the 4-position of the cyclohexyl ring. Benzamides such as sulpiride comprise a well-known class of DA antagonists. But when applied in a somewhat different context, benzamide functionality attached via the cyclohexyl tether to an aryl amine led to DA partial agonists. A series of analogues probed the importance of distance and stereochemistry between the pyridinyl piperazine and the benzamide functionalities and some analogues were found to have potent activity as DA autoreceptor agonists.[77]

The distance between the piperazine and benzamide groups was varied by placing a 0–3-carbon atom spacer between the cyclohexyl and piperazine rings. The effect of the relative stereochemistry between these functional groups was also determined. The DA D2 receptor affinity of this series was measured using rat striatal tissue with [^{3}H]spiperone as the radioligand and, in certain cases, the agonist radioligand [^{3}H]N-propyl apomorphine. As can be seen in Table 11, in the case of 0-, 1-,

Table 11. Cyclohexylbenzamide DA D2 Agonists

Compound	n	Isomer	[^{3}H]Spiperone Binding IC_{50}, nM	[^{3}H]NPA[a] Binding IC_{50}, nM	% Reversal of DOPA Synthesis After GBL	Inhibition of Mouse Locomotor Activity ED_{50} mg/kg ip
55	0	trans	2800	110	65	0.36
56	0	cis	>10000			>30
57	1	trans	1240	7.6	36	0.71
58	1	cis	1800			3.5
59	2	trans	443	19	87	0.16
60	2	cis	3280		36	4.5
61	3	trans	1230	24	37	3.6
62	3	cis	1050	21	61	2.7

Note: [a]NPA = N-propyl apomorphine.

and 2-carbon spacers (compounds **55–60**), the *cis*-isomers had weaker DA D2 receptor affinity than the *trans*-isomers. In each case, the *trans*-isomer was significantly more potent at inhibiting locomotor activity after ip administration in the mouse. In the case of the 3-carbon linker activity was about equal for the *cis*- and *trans*-analogues. All of these compounds had relatively low receptor affinity as assessed using [³H]spiperone as the radioligand. However, this series appeared to have much greater affinity when the agonist radioligand [³H]N-propyl apomorphine was used. These data suggested that this series was interacting predominantly with the high-affinity (agonist) form of the receptor as one would expect for an agonist or partial agonist. The ability of the compounds to reverse the GBL-induced increase in dopamine synthesis in the rat striatum was used as an in vivo measure of agonist activity. Despite only modest differences in binding affinity, compound **59** was most active in this regard (87% inhibition at 20 mg/kg ip). Compound **59** was also most potent at inhibiting locomotor activity in the mouse after ip administration as shown in Table 11. This compound had a *trans* orientation about the cyclohexyl ring between the benzamide and the ethylene linker bearing the pyridyl piperazine functionality.

We considered which conformations of compound **59** might be responsible for the DA agonist activity. Molecular modeling studies carried out using the Macromodel program confirmed, as expected, that in all low-energy conformations of the molecule the ethylene chain and the amide nitrogen were organized in a diequatorial manner. Further analysis revealed that there were a number of conformations of relatively low energy that the ethylene chain could adopt. These various dihedral angles resulted in a variety of potential orientations between the pyridyl–piperazine ring and the cyclohexane portion of the molecule. In this case molecular modeling studies could not offer a conclusive answer to the relationship between the benzamide and aryl piperazine functionality found in the molecule.

B. Rigid Bicyclics

To further pursue this question, compounds were prepared wherein the ethylene side chain was incorporated into a second cyclohexane ring (Figure 5).[78] Such 6,6-bicyclic systems, known as decalin ring systems, can exist having either a *trans* or *cis* ring juncture. Compounds having a *trans* ring juncture have a high degree of conformational rigidity. Thus if the benzamide and pryidinyl piperazine functionality could be placed

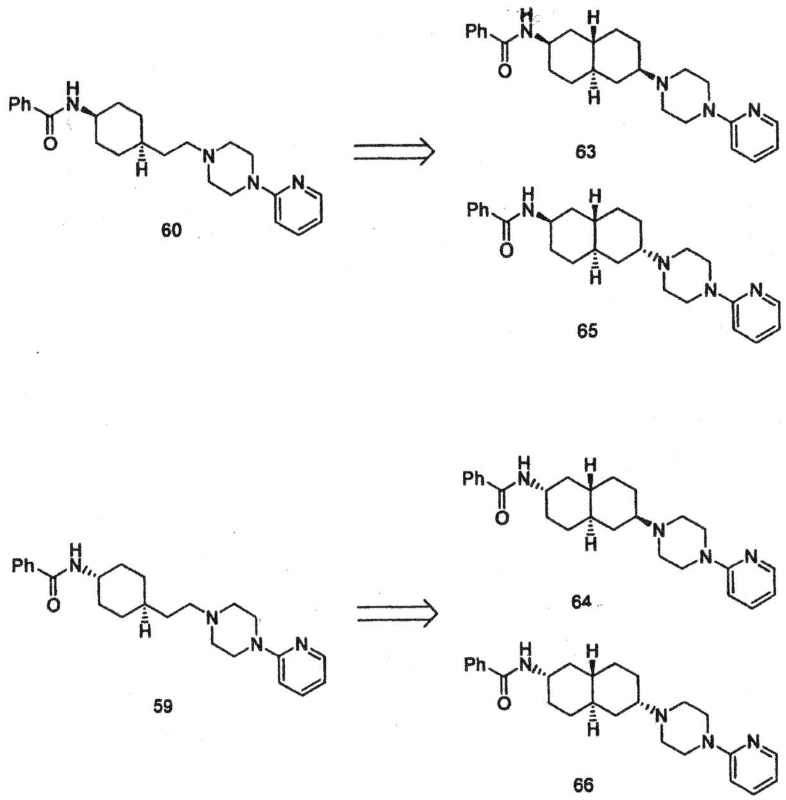

Figure 5. Rigid bicyclic benzamides.

around this ring system with known relative stereochemistry, specific conformations of **59** could be approximated. The four possible decalin diastereomers (**63** to **66**) were prepared and relative stereochemistry determined by proton and ^{13}C NMR measurements.[78] For compounds **63** and **64** the pyridyl piperazine is held in the equatorial position. In compound **63** the pyridyl piperazine is held in an axial orientation. In compound **64** the benzamide is held in an equatorial orientation. Compounds **65** and **66** have the pyridyl piperazine functional group oriented in an axial direction. Compounds **63** and **65** have the benzamide held in an axial orientation and are likely to represent the conformations of the *cis*-isomer **60** where the benzamide functionality is most likely in an axial orientation. Based on the earlier observations that *trans*-isomers such as

59 had greater receptor affinity than the *cis*-isomer 60 it was expected that 64 and 66 would have greater receptor affinity than the isomers 63 and 65.

The binding properties and functional activity of 59 and the rigid analogues 63–66 were studied for DA D2 receptors and summarized in Table 12. Compound 64 exhibited tight binding affinity and was a full agonist in functional tests with a potency and intrinsic activity similar to that observed for 59. The other isomers were significantly weaker in these in vitro assays. Compound 63, like compound 64, had an equatorial orientation of the pyridyl piperazine functionality but, unlike 64, the benzamide functionality of 63 was oriented in the axial position. This resulted in compound 63 having significantly weaker binding interactions and functional activity. Isomers 65 and 66 had the pyridinyl piperazine held in the axial orientation and both compounds had greatly reduced affinity for the DA D2 receptor and both compounds completely lacked activity in the functional assay.

Of the four conformationally locked analogues related to compound 59, only the isomer 64 with the benzamide and aryl piperazine functionalities both in an equatorial conformation had comparable binding and functional activity. This was consistent with the notion that 59 was bound to the DA D2 receptor in an extended conformation. Energy-minimized structures of 59 and 63–66 were modeled to determine the distance between the aryl centroid of the benzamide and the basic piperazine nitrogen atom (Figure 6). For 64 the distance was 11.8 Å and corresponds to the distance observed in the extended conformation of compound 59,

Table 12. In Vitro Properties of Rigid Analogues

Compound	[³H]Spiperone Binding (IC$_{50}$, nM)	Functional Activity[a] EC$_{50}$ (nM)	Intrinsic Activity[b]
59	443	4.9	0.83
63	1100	240	0.63
64	72.7	10.4	0.98
65	7100	>1000	
66	5800	>1000	
quinpirole		7.6	1.0

Notes: [a]Inhibition of forskolin-stimulated cAMP accumulation in GH4C1 cells transfected with the DA D2 receptor.

[b]Intrinsic activity relative to quinpirole.

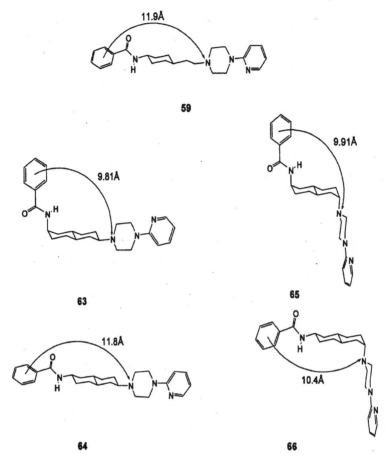

Figure 6. Energy-minimized structures of benzamide DA agonists.

11.9 Å. The arylcentroid-nitrogen distance of the other three isomers was 1.5–2 Å shorter, again suggesting that this distance is critical for receptor recognition. Although the rigid compounds were useful in vitro tools, their solubility properties in aqueous media were such that they were not ideal for in vivo studies.

C. D3 Preferring Compounds

From molecular biology studies it has been determined that the population of DA receptors labeled by classic "D2" radioligands is in fact composed of three distinct receptor subtypes labeled D2, D3, and

D4.[16–18] Although the exact role of the D3 receptor in the CNS is not clear there is some evidence suggesting that D3 receptors may play a role in regulating dopamine synthesis.[79,80] Binding studies in our laboratories determined that certain members of the cyclohexyl amide series had preferential affinity for the novel DA D3 receptor subtype compared to the more abundant DA D2 receptor. Cyclohexyl-thiopheneamide 67 was found through high volume screening to have good affinity ($K_i = 4.3$ nM) and moderate selectivity (16-fold) for the D3 versus the D2 receptor.[81] SAR studies involving the aryl group adjacent to the piperazine (Table 13) revealed that replacement of the pyridine ring with a phenyl (compound 68) led to a loss of D3 selectivity. However introduction of electron-withdrawing 2,3-dichloro substituents on the phenyl group led to somewhat greater selectivity. Compound 72 having chloro substitution at the 2- and 3-positions of the aromatic ring attached to the piperazine nitrogen had picomolar affinity for the D3 receptor and 30-fold selectiv-

Table 13. D3 Preferring Cyclohexylbenzamides

No.	R_1	R_2	Ar	Stereo-chemistry	$D2^a$ (K_i nM)	$D3^b$ (K_i nM)	D2/D3
67	2-thienyl	H	2-pyridyl	trans	71	4.3	16.5
68	2-thienyl	H	phenyl	trans	3.4	0.8	4
69	2-thienyl	H	2-Cl phenyl	trans	1.0	0.5	2
70	2-thienyl	H	3-Cl phenyl	trans	6.5	2.0	3
71	2-thienyl	H	4-Cl phenyl	trans	34	8.0	4
72	2-thienyl	H	2,3-diCl phenyl	trans	0.6	0.02	30
73	2-thienyl	H	2,3-diCl phenyl	cis	18.0	1.8	10
74	2-furanyl	H	2,3-diCl phenyl	trans	3.0	1.4	2
75	3-thienyl	H	2,3-diCl phenyl	trans	5.0	4.7	1
76	2-thienyl	Me	2,3-diCl phenyl	trans	16.5	4.2	4
77	methyl	H	phenyl	trans	10	2.5	4
78	cyclohexyl	H	phenyl	trans	38	0.14	270
79	cyclohexyl	H	2,3-diCl phenyl	trans	7.4	4.5	2

Notes: [a]Binding studies were carried out in CHO-K1 cells transfected with the long form of the human D2 receptor using the agonist radioligand [^{3}H]N-0437.

[b]Binding studies were carried out in CHO-K1 cells transfected with the human DA D3 receptor using [^{3}H]spiperone as the ligand.

ity. The corresponding *cis*-isomer **73** was weaker and less selective than the *trans*-isomer **72**. Despite its extreme potency for D3 receptors, compound **72** still retained significant affinity for D2. The amide N–H appears to play an important role in the affinity of **72** for both the D3 and D2 receptors as the *N*-methyl analogue **76** showed greatly reduced affinity for both receptors. A further study of various amides demonstrated that substitution of a furanyl as in compound **74** led to relatively weaker and much less selective D3 ligands.

Compound **77**, a simple acetamide analogue with an unsubstituted phenyl group, had nM D3 affinity; moreover, replacement of the acetamide with a cyclohexylcarboxamide led to compound **78** with sub-nM potency for the D3 receptor and over 270-fold selectivity for D3 vs D2 receptors. Substitution of the phenyl ring with 2,3-dichloro substitution resulted in compound **79** with decreased D3 affinity and selectivity.

The two most potent and selective compounds at the D3 receptor, **72** and **78,** were characterized for their intrinsic activity at the cloned human D2 and D3 receptors by using stimulation of [^{3}H]thymidine uptake as a measure of in vitro functional activity (Table 14). Both compounds behaved as partial agonists at D2 receptors. Compound **78** was approximately 20-fold more potent as a partial agonist at D3 receptors, while **72** was an antagonist at the D3 receptor. Despite their apparent selectivities in the binding assay, both compounds retained significant functional activity at D2 receptors even at low nM concentrations. The fact that they are both partial agonists at DA D2 receptors may explain why both compounds caused decreases in limbic DOPA synthesis after GBL administration. If D3 receptors played a larger role than D2 receptors in controlling DA synthesis and release, one would expect to see an increase

Table 14. Functional Activity at D2 and D3 Receptors

Assay	72	78
D2 Functional assay	43%	73%
(% IA, EC_{50} or IC_{50} in nM)	EC_{50} 5.1 nM	EC_{50} 4.9 nM
D3 Functional assay	0%	44%
(% IA, EC_{50} or IC_{50} in nM)	IC_{50} 0.99 nM	EC_{50} 0.28 nM
% Reversal of DOPA accum. after GBL (@10 mg/kg ip)	75	100

in DOPA levels after administration of a D3 antagonist such as **72**. Since a decrease in DOPA levels was observed with **72**, it is likely these effects are mediated through the D2 receptor.

V. HETEROCYCLIC REPLACEMENTS FOR THE AMIDES

A. Indoles

While the cyclohexyl amide series displayed attractive in vitro profiles and desired in vivo activity profiles after ip administration, these compounds often lacked significant activity in behavioral tests after oral administration. A plausible explanation is that poor metabolic stability led to the rapid formation of less active cyclohexyl amine analogues by amide hydrolysis. To prevent this from occurring, we sought to replace the amide functional group with various heterocycles. Mindful of our earlier success in preparing DA partial agonists with either phenyl or hydrogen bond-donating amide functionality appended to the cyclohexyl ring, exploration of indole functionality containing both of these pharmacophores appeared to be an avenue worth pursuing.[82]

Compounds were evaluated for their ability to bind to the dopamine D2 receptor in vitro and inhibit locomotor activity in mice after ip administration (Table 15). As had been previously observed in the amide series, increasing chain length between the cyclohexyl and piperazine rings led to increases in affinity for the DA D2 receptor as measured by [³H]spiperone binding in rat striatal membranes. In this case the increase was somewhat larger. In analogy with the amide case the *trans*-isomers were 1.5–5-fold more potent at the D2 receptor than the corresponding *cis*-isomers. In a series of *cis*- and *trans*-isomers having 0–2 carbon atoms as the spacer between the cyclohexyl and piperazine (compounds **80–85**), compound **85** with a *trans* 2-carbon linker between the indole and pyridinyl piperazine groups was most active. Replacement of the 1-(2-pyridyl)piperazine functionality with a variety of other arylamino groups (compounds **86–91**) generally led to decreased affinity for the DA D2 receptor. The lone exception was the phenyl tetrahydropyridine analogue **87** which was approximately threefold more potent for the DA D2 receptor, but this analogue was inactive in vivo. Changes to the indole portion of the molecule also led to a modulation of receptor binding affinity. Compound **92** having an electron-withdrawing fluorine substituent in the 5-position of the indole ring suffered a twofold decrease in binding affinity. However, analogues **93** and **94** having electron-donating

Table 15. Cyclohexyl Indole SAR

No.	n	Stereochemistry	NR₁R₂	X	Y	[³H]Spiperone Binding IC_{50}, nM	Inhibition of Mouse Locomotor Activity ED_{50} mg/kg ip	Inhibition of Rat Locomotor Activity ED_{50} mg/kg po	% Reversal of Striatal DOPA Synthesis After GBL
80	0	cis	piperazinyl-pyrimidine	H	H	2596	>30	NT[a]	NT
81	0	trans	piperazinyl-pyrimidine	H	H	457	1.2 (0.7; 2.2)	13.7 (7.5; 25.1)	38 ± 5.9
82	1	cis	piperazinyl-pyrimidine	H	H	122	1.2 (0.6; 2.2)	weak[b]	NT
83	1	trans	piperazinyl-pyrimidine	H	H	85	2.3 (1.4; 3.7)	8.8 (6.9; 11.1)	81 ± 5.6
84	2	cis	piperazinyl-pyrimidine	H	H	139	4.7 (3.5; 6.5)	12.2 (7.8; 19.2)	60 ± 9.9
85	2	trans	piperazinyl-pyrimidine	H	H	50	1.4 (0.7; 2.6)	4.8 (3.0; 7.6)	44 ± 4.4
86	2	cis	phenylpiperidine	H	H	28	10.5 (8.8; 12.7)	NT	5 ± 5
87	2	trans	phenylpiperidine	H	H	8.7	7.2 (2.4; 21.1)	>30	INC[c]

143

Table 15. Continued

No.	n	Stereo-chemistry	NR_1R_2	X	Y	$[^3H]$Spiperone Binding IC_{50}, nM	Inhibition of Mouse Locomotor Activity ED_{50} mg/kg ip	Inhibition of Rat Locomotor Activity ED_{50} mg/kg po	% Reversal of Striatal DOPA Synthesis After GBL
88	2	trans	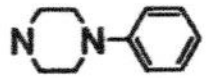 (structure)	H	H	124	13.9 (5.7; 34.0)	NT	INC
89	2	trans	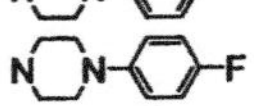 (structure)	H	H	645	>30	NT	NT
90	2	trans	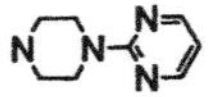 (structure)	H	H	161	3.3(2.2; 5.0)	18.6 (10.7; 32.4)	34 ± 13
91	2	trans	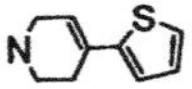 (structure)	H	H	630	11.6 (4.6; 29.5)	NT	NT
92	2	trans	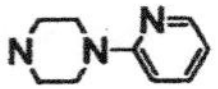 (structure)	F	H	110	9.0 (4.9; 16.4)	Stim	100 ± 13
93	2	trans	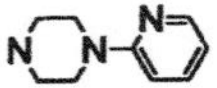 (structure)	OMe	H	29	2.2 (1.8; 2.7)	>30	88 ± 1.3
94	2	trans	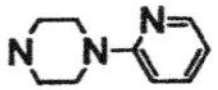 (structure)	OH	H	8.6	0.6 (0.4; 0.9)	>30	25 ± 11
95	2	trans	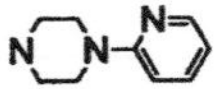 (structure)	H	Me	83	>30	NT	NT
roxindole						14	0.25 (0.12; 0.52)	47.6 (15.7; 144.2)	70 ± 2.2

Notes: [a]NT = not tested.

[b]Inhibition of locomotor activity never reached 50%.

[c]INC = increase in DOPA levels.

144

methoxy and hydroxy substituents in the 5-position had increased receptor binding affinity. Substitution of the indole N–H with a *N*-methyl (compound **95**) resulted in only a slight decrease in binding affinity and a complete loss of in vivo activity.

The in vivo activity of these compounds diverged somewhat from their D2 receptor binding affinity. This is not surprising as a number of factors other than receptor binding affinity (absorption, metabolic half-life, blood–brain barrier penetration) all contribute to observed whole animal pharmacology. Out of this series, compound **85** was chosen for further evaluation because it had the best overall in vivo profile (Table 16). It was among the most potent at inhibiting mouse locomotor activity after ip administration and was the most potent in a similar rat paradigm after oral administration. In the rat paradigm, **85** was approximately 10-fold more potent than roxindole. The compound also decreased DA synthesis in the rat. Although some compounds caused larger decreases at similar doses (i.e. **92** and **93**), they were less active at inhibiting rat locomotor activity or caused stimulation.

Despite its good overall in vivo profile, compound **85** showed only 50 nM potency in the [^{3}H]spiperone binding assay. However when binding

Table 16. Compound **85** Profile

Test	85
DA D2 receptor binding [^{3}H]NPA (K_i, nM)	2.6
Inhibition of 3H Thymidine uptake	
Intrinsic activity DA D2 receptor[a]	72%
EC$_{50}$ (nM)	2.8
DA D3 receptor binding [^{3}H]spiperone (K_i, nM)	29
DA D4.2 receptor binding [^{3}H]spiperone (K_i, nM)	73
DOPA accumulation in rats after GBL[b] (ED$_{50}$, mg/kg ip)	6.6
Decrease of rat striatal dopamine overflow (10 mg/kg, ip)[c]	33%
Decrease in DA neuronal firing rate in rats (2.5 mg/kg ip)	100%
Inhibition of APO-induced climbing in mice (ED50, mg/kg/ip)	>30
Stereotypy in rats (ED$_{50}$, mg/kg, ip)	>24
Decrease in squirrel monkey striatal dopamine overflow (3 mg/kg ip)[d]	38%

Notes: [a]Data expressed relative to quinpirole (100%).

    [b]Graphically determined ED50 value, defining 50% reversal of the increase in DOPA accumulation induced by GBL.

    [c]Measured via *in vivo* microdialysis ($n = 4$).

    [d]Measured via *in vivo* microdialysis ($n = 1$).

studies were carried out using [³H]-NPA, a ligand that labels only the high-affinity state of the receptor, compound **85** had an approximately 25-fold lower K_i (Table 16). This suggests that **85** binds with greater affinity to the agonist form of the DA D2 receptor labeled by the [³H]-NPA. The compound had 10–20 fold weaker affinity at the DA D3 and D4 receptor subtypes, suggesting most of the effects observed occurred through DA D2 receptors. Studies of the functional effects of the compound on human DA D2 receptors were carried out in a CHO-P5 cell line into which the long form of the receptor had been cloned. DA agonists stimulate [³H]thymidine uptake in these cells. Compound **85** was classified as a partial agonist in this assay having intrinsic agonist efficacy of 72% compared to the full DA D2 agonist quinpirole. The EC_{50} of this effect was 2.8 nM, similar to that of the observed agonist receptor binding K_i.

Studies in rodent models suggested a DA D2 autoreceptor mechanism of action for **85**. In electrophysiological studies in anesthetized rats, compound **85** was able to completely block firing of substantia nigra DA neurons, an effect believed to involve activation of presynaptic DA D2 autoreceptors. Inhibition of brain dopamine synthesis in rats as measured by decreases in GBL induced DOPA levels, as well as decreases in striatal DA levels in in vivo microdialysis studies, were also consistent with a DA autoreceptor mechanism of action. Similar results were measured in in vivo microdialysis studies carried out in a squirrel monkey, suggesting that the compound was also acting as a DA partial agonist. Reductions in the DA levels of the caudate putamen of a monkey were observed after ip administration of **85**.

Earlier, workers at E. Merck had prepared indole alkyl amines such as roxindole with simple alkyl spacers between the indolyl and tetrahydropyridine functionality.[62] While these compounds appeared to have a profile consistent with selective autoreceptor activation in rodents, roxindole was inactive in the Sidman primate model of antipsychotic efficacy in our laboratories.[83] The findings in primates corresponded to results in the clinic that showed roxindole to be ineffective for the treatment of the positive symptoms of schizophrenia.[84] These studies bolstered the argument that activity in this primate model of antipsychotic efficacy might serve as a useful marker for the identification of compounds potentially active against schizophrenia. Compound **85** potently inhibited this behavioral response with an ED_{50} of 1 mg/kg po (see Table 10). Taken together, the in vitro and in vivo profiles of **85** demonstrate that DA D2 partial agonist activity in vitro can translate into measurable decreases in DA synthesis in vivo and activity in behavioral tests predictive of antipsychotic efficacy.

The propensity for causing EPS was assessed in haloperidol-sensitized monkeys (Table 10). Studies of compound **85** revealed that despite its properties suggesting selective autoreceptor activation, the compound caused strong extrapyramidal symptoms at doses only 10 times above its ED_{50} in the Sidman avoidance paradigm. Agents such as olanzapine and rispiridone which in clinical trials have been shown to have a reduced risk of extrapyramidal side effects show similar activity in this primate measure of side effect liability.

B. Aminopyrimidines

An alternative method for mimicking the amide functionality of compounds such as **68** was to use a heterocycle such as a pyrimidine in place of the amide carbonyl. Preparation of compounds of this type led to the discovery of an interesting series of aminopyrimidines that, in addition to having partial agonist activity at DA D2 receptors, also possessed partial agonist properties at 5-HT1A receptors.[85] Such a profile is attractive as 5-HT1A agonists and partial agonists have been shown to block the cataleptic effects caused by blockade of dopaminergic neurotransmission.[86–89] On this basis it was postulated that a compound having a mixed DA D2 and 5-HT1A partial agonist profile might have utility as an antipsychotic agent.

As in the analogous amide series the effect of stereochemistry and distance of the cyclohexane ring from the arylpiperazine ring was examined. In these studies distinct structure–activity relationships for the DA D2 and 5-HT1A receptors were observed (Table 17). Direct attachment of the aryl piperazine to the cyclohexane as in compounds **97** and **98** led to compounds with only weak binding affinity for the D2 receptor while the *trans*-isomer had moderate affinity for the 5-HT1A receptor. Increasing this distance by one carbon atom resulted in an increase in D2 affinity while 5-HT1A affinity remained about the same. Introduction of a 2-carbon spacer between the cyclohexane and piperazine rings having a *cis* orientation relative to the aminopyrimidine ring (compound **100**) resulted in good potency at the 5-HT1A receptor and moderate affinity at the D2 receptor. Compound **101** having a 2-carbon spacer and *trans* orientation relative to the aminopyrimidine group had good potency for both D2 and 5-HT1A receptors. Addition of a third carbon into the spacer (compound **102**) resulted in little decrease in DA D2 potency but an

Table 17. Amino-Heterocycle Structure–Activity Relationships

	R	n	Stereochemistry	Binding (K_i, nM)		
				D2[a]	D3[b]	5-HT1A[c]
97	(pyrimidin-2-yl-NH)	0	cis	160	550	470
98	(pyrimidin-2-yl-NH)	0	trans	320	610	83
99	(pyrimidin-2-yl-NH)	1	trans	5.7	89	41
100	(pyrimidin-2-yl-NH)	2	cis	28	29	1.9
101	(pyrimidin-2-yl-NH)	2	trans	5.2	13.7	3.5
102	(pyrimidin-2-yl-NH)	3	trans	7.4	54	16
103	(pyrimidin-2-yl-N-Me)	2	trans	4.8	24	3.7
104	(quinazolin-2-yl-NH)	2	trans	3.2	0.2	74
68	(thiophene-2-carboxamido)	2	trans	3.4	0.8	57
78	(cyclohexanecarboxamido)	2	trans	38	0.14	5300

Notes: [a]Binding studies were carried out in CHO-K1 cells transfected with the long form of the human D2 receptor using the agonist radioligand [^{3}H]N-0437.

[b]Binding studies were carried out in CHO-K1 cells transfected with the human DA D3 receptor using [^{3}H]spiperone as the ligand.

[c]Binding studies were carried out in a rat hippocampal membrane preparation using [^{3}H]8-OH-DPAT as the radio ligand.

approximate threefold drop in 5-HT1A potency. The hydrogen bond-donating properties of the aminopyrimidine group did not appear to be important for the observed receptor affinity since the *N*-methyl analogue **103** also had similar receptor affinity. The 5-HT1A receptor appeared to be very sensitive to substitution on the pyrimidine ring as the quinazoline analogue **104** had a greater than 10-fold decrease in affinity at this receptor but its D2 affinity was unchanged. Amides **68** and **78** were also significantly weaker at the 5-HT1A receptor, suggesting amino pyrimidine functionality is very important for recognition of the series by this serotonergic receptor.

The phenyl tetrahydropyridine derivative **105** had similar affinity to the phenyl piperazine derivative **101**; however, this analogue was deemed less interesting because of concerns about metabolism raised by our earlier efforts with the tetrahydropyridine derivative CI-1007. The tetrahydroisoquinoline and phenethyl amine derivatives **106** and **107**, respectively, showed reduced affinity for both receptors. The amino derivative **108**, which lacked adjacent aromatic functionality, had extremely weak affinity for DA D2 receptors.

With these initial results in hand attention was focused on a series of substituted phenyl and heterocyclic piperazines compounds **109** to **119** (Table 18). The dichloro analogue **109** had similar affinity for D2 receptors but a nearly 25-fold decrease in affinity for 5-HT1A receptors. The 2-methoxyphenyl analogue **110** retained potent affinity for both D2 and 5-HT1A receptors, while the 4-methoxyphenyl analogue **111** was much less active at both receptor subtypes. Compound **112** having a 3-trifluoromethyl substituent on the aromatic ring led to a fourfold decrease in activity, while introduction of a fluoro substituent (compound **113**) led to a similar binding profile to the unsubstituted phenyl piperazine **101**. A number of heterocyclic replacements for the phenyl ring of **101** were also studied. The 2-pyridyl analogue **114** showed a twofold decrease in DA D2 receptor binding affinity but had slightly better potency at the 5-HT1A receptor while the 3-pyridyl analogue **115** showed five- and threefold decreases at the DA D2 and 5-HT1A receptors, respectively. The 4-pyridyl analogue **116** had large decreases in affinity for both receptors. The 2-pyrimidinyl analogue **117** had good potency for 5-HT1A receptors but was significantly weaker (ca. 50-fold) than the parent compound **101** at DA D2 receptors. The 2-pyrazinyl and 2-thiazolyl analogues **118** and **119**, respectively, also had larger decreases at the DA D2 receptor in comparison to their loss in 5-HT1A affinity. Overall this study suggested that heterocyclic replacements for

Table 18. Aminopyrimidine SAR

	NR$_1$R$_2$	Receptor Binding (K$_i$, nM)		
		D2[a]	D3[b]	5-HT1A[c]
101		5.0	14	3.5
105		4.7	3.7	3.9
106		110	40	44
107		39	NT[d]	91
108		2400	800	ND
109		11	5	76
110		3	11	0.5
111		390	52	160
112		6	32	13
113		8.5	48	4.5
114		12	26	2.2
115		25	4.7	9.5
116		4800	2300	730
117		240	590	8.6
118		180	210	29
119		46	ND	10

Notes: [a]Binding studies were carried out in CHO-K1 cells transfected with the long form of the human D2 receptor using the agonist radioligand [^{3}H]N-0437.

[b]Binding studies were carried out in CHO-K1 cells transfected with the human DA D3 receptor using [^{3}H]spiperone as the ligand.

[c]Binding studies were carried out in a rat hippocampal membrane preparation using [^{3}H]8-OH-DPAT as the radio ligand.

[d]NT = not tested.

the phenyl ring of **101** had smaller effects on 5-HT1A binding than on DA D2 receptor affinity.

Based upon their binding characteristics four compounds were selected for further evaluation. Intrinsic agonist efficacy at human DA D2 receptors transfected into a CHO p-5 cell line were studied by measuring stimulation of [³H]thymidine uptake in these cells compared to the agonist quinpirole. Compounds with low intrinsic activity were studied for their ability to block effects of the nonspecific DA agonist quinpirole (Table 19). The unsubstituted phenyl piperazine **101** behaved as a partial agonist in this test system with 60% intrinsic activity, while the 2-methoxy and 4-fluoro analogues **110** and **113**, respectively, both behaved as DA D2 antagonists in this in vitro assay. The 2-pyridyl analogue **114** behaved as a partial agonist with lower intrinsic activity than compound **101**. Intrinsic efficacy at DA D2 and 5-HT1A receptors was assessed in vivo by measuring changes in limbic DOPA and 5-hydroxy tryptamine (5-HTP) levels, respectively. The partial DA agonist **101** caused decreases in both neurotransmitter metabolites, indicating presynaptic activity at DA autoreceptors and somatodendritic 5-HT1A receptors. The DA D2 antagonists **110** and **113** as expected caused increases in DOPA levels as well as in 5-HTP levels, indicating antagonist activity at both receptors in vivo. It has been generally appreciated that a 2-methoxy substituent on an aryl piperazine ring tends to confer antagonist activity at dopaminergic and serotonergic receptors. Similarly in this instance, a 4-fluoro substituent resulted in an antagonist profile at both DA D2 and

Table 19. Functional Activity at D2 and 5-HT1A Receptors

No.	Inhibition of [³H] Thymidine Uptake (% Intrinsic Activity, EC_{50} or IC_{50} in nM)[a]	% of Control in Limbic DOPA Levels 10 mg/kg ip[b]	% of Control in Limbic 5-HTP Levels 10 mg/kg ip[b]
101	60% EC_{50} = 29	67 ± 4	57 ± 5
110	8% IC_{50} = 9.9	173 ± 3	128 ± 7
113	5% IC_{50} = 79	142 ± 9	143 ± 6
114	41%, EC_{50} = 28	159 ± 3	53 ± 3

Notes: [a]Effects measured in CHO p-5 cells transfected with the h-D2 receptor. Intrinsic activity measured relative to the full agonist quinpirole. In cases with less than 20% increase, blockade of quinpirole effects were measured to determine IC_{50}.

[b]Test compounds were administered ip (10 mg/kg) 30 min before the decarboxylase inhibitor NSD 1015 (100 mg/kg ip) and animals were sacrificed 30 min after the NSD 1–15. Each value is a mean of 4–8 animals and is expressed as a percent of control values 974 ± 16 (ng/g ± SEM) and 469 ± 5 (mg/g ± SEM) for mesolimbic DOPA and 5-HTP levels, respectively.

5-HT1A receptors. The three phenyl piperazine compounds appeared to have roughly similar levels of intrinsic activity in vivo at both DA D2 and 5-HT1A receptors. However, in the case of the pyridyl piperzine analogue **114** antagonist effects were observed at DA D2 receptors as indicated by an increase in DOPA levels. Although this compound displayed weak partial agonist activity, its level of intrinsic activity is apparently not strong enough to exhibit D2 receptor agonist actions in vivo. In contrast to this antagonist activity at DA D2 receptors, partial agonist activity was observed at 5-HT1A receptors as evidenced by decreases in 5-HTP levels.

Taken together, the data from binding, in vitro functional activity, and in vivo neurochemical assays suggest that the amino pyrimidine and aryl piperazine play different roles at each receptor. For the DA D2 receptor the nature of the aryl group on the piperazine ring appears to play a dominant role in receptor recognition. The substitution pattern and nature of the aryl ring system also appears to play a large role in determining intrinsic activity at the DA D2 receptor. A different pattern emerges for the 5-HT1A receptor. In this case the aminopyrimidine portion of the molecule appears to play a more dominant role in recognition of compounds by this receptor. The nature of the aryl piperzine group, however, is important in determining what the intrinsic activity the 5-HT1A receptor ligand may be. It is interesting to note that in one instance in this series we were able to observe in vivo neurochemical changes indicating agonist activity at 5-HT1A receptors but antagonist activity at the DA D2 receptors. This result suggests different structural requirements for agonist activity at these two presynaptic neurotransmitter receptors.

The neurochemical effects of **101** were studied using whole brain microdialysis. In agreement with the synthesis studies, administration of the compound at 10 mg/kg ip caused significant decreases in striatal levels of DA and the 5-HT metabolite 5-HIAA (Table 20). Compound **101** was also studied for activity in a variety of behavioral paradigms predictive of antipsychotic efficacy. It effectively decreased locomotor activity in both mice after ip administration and rats after oral administration. In the monkey Sidman avoidance model, compound **101** had both good efficacy and excellent oral potency offering a proof of concept that compounds with partial agonist activity at both DA D2 and 5-HT1A receptors would be potently efficacious antipsychotic agents (see Table 10). The EPS liability of this compound was assessed in cebus monkeys. No effects were observed at the ED_{50} dose but were observed in 3/3

Table 20. Neurochemical and Behavioral Effects of Compound **101**

Test	Result
% Decrease of rat striatal dopamine overflow (10 mg/kg, ip)[a,b]	50 (36; 64)
% Decrease of rat striatal 5-HIAA overflow (10 mg/kg, ip)[a,b]	42 (28; 56)
Inhibition of spontaneous locomotor activity in mice (ED50 ip)[b]	0.4 (0.3; 0.5)
Inhibition of spontaneous locomotor activity in rats (ED50 po)[b]	4.6 (3.2; 6.4)
Inhibition of Sidman avoidance in squirrel monkey (ED$_{50}$, mg/kg po)[b]	0.28 (0.02; 0.34)

Notes: [a] Measured via *in vivo* microdialysis ($n = 4$).
[b] 95% confidence limits in parenthesis.

monkeys at 5 times this dose. Development of this compound was terminated after GI toxicity was observed.

VI. CONCLUSIONS

Several conclusions can be drawn from these studies. Structure–activity relationships of several series of DA agonists and partial agonists have been studied. Rigid analogues were prepared that suggest that indeed these compounds exert their agonist effects at the DA D2 receptor in an extended conformation. This class of molecules is distinct from the earlier class of "classic" DA agonists which has elements that mimic the catechol and amine functions found in DA. Thus two fundamentally different structures can act as agonists or partial agonists at the DA D2 receptor. It is possible that these various structural types interact with different active conformations of the receptor. Recent site-directed mutagenesis work on DA D2 receptors support the notion of more than one active conformation for this receptor. Within this new class of DA agonists we have shown it possible to tune intrinsic activity such that compounds possessing high, medium, or low levels of agonist activity can be produced. This can be accomplished by regio- and stereochemical manipulations as well as by changes in substitution patterns.

On a pharmacological level we have shown it is possible to create compounds with the appropriate level of intrinsic activity to selectively stimulate presynaptic DA D2 autoreceptors. This results in compounds that measurably decrease DA synthesis and release in both rodents and primates. Such compounds show a lack of postsynaptic receptor stimulation. Behaviorally these compounds inhibit locomotor activity in mice and rats. In monkeys these compounds demonstrated potent activity in the Sidman paradigm after oral administration (see Table 10). This test is predictive of antipsychotic efficacy, suggesting these partial agonists would be effective in the clinic. CI-1007 was shown to have no propensity for causing extrapyramidal side effects at the Sidman ED_{50} dose. Multiples up to 9 times its ED_{50} caused EPS in only half the monkeys studied. Compound **85** was slightly less potent in the Sidman paradigm but had similar EPS when tested at 10 mg/kg po in the cebus monkey. The mixed D2/5-HT1A partial agonist **101** was efficacious in the Sidman paradigm and did not have EPS at its ED_{50} concentration. However, a dose of 5 times the ED_{50} caused significant EPS in the haloperidol-sensitized monkeys. The results suggest that DA partial agonists with low enough intrinsic activity to inhibit primate Sidman avoidance responding will also cause EPS at higher doses in haloperidol-sensitized monkeys. However, this model may overrepresent the actual risk of the development of EPS by patients not currently sensitized to haloperidol. The fact that an increased separation was observed between Sidman ED_{50} and minimum doses causing EPS doses suggest that these compounds may not have extrapyramidal liability in the clinic. To date, no DA D2 autoreceptor agonist has been successfully developed as an antipsychotic agent leaving this hypothesis as yet unproven in the clinic.

ACKNOWLEDGMENTS

The author acknowledges colleagues in the Parke-Davis Neuroscience Therapeutics Departments who were instrumental in the electrophysiological, neurochemical, and behavioral studies including Thomas Heffner, Thomas Pugsley, and Leonard Meltzer. The author would like to acknowledge colleagues in the Chemistry Department who designed and prepared these compounds under the direction of Lawrence Wise. These include Juan Jaen, Brad Caprathe, Jon Wright, William Smith III, Thomas Belliotti, Thomas Capiris, Shelly Glase, Dennis Downing, and Suzanne Kesten. The author wishes to dedicate this review to the memory of the late Stephen Johnson who was instrumental in the synthesis of many of these compounds. His excellence in synthetic chemistry and friendship will be greatly missed.

REFERENCES

1. Seeman, P. *Synapse* **1987**, *1*, 133–152.
2. Carlsson, A.; Hansson, L. O.; Waters, N.; Carlsson, M. L. *Life Sci. (Washington, DC)* **1997**, *61*, 75–94.

3. Farde, L. *Schizophr. Res.* **1997**, *28*, 157–162.
4. Ackenheil, M.; Hippius, H. *Psychopharmacology (N.Y.)* **1977**, *2*, 923–956.
5. Schmutz, J.; Eichenberger, E. *Chron. Drug Discovery* **1982**, *1*, 39–59.
6. Leysen, J. E.; Gommeren, W.; Eens, A.; de Chaffoy de Courcelles, D.; Stoof, J. C.; Janssen, P. A. *J. Pharmacol. Exp. Ther.* **1988**, *247*, 661–670.
7. Skarsfeldt, T.; Perregaard, J. *Eur. J. Pharmacol.* **1990**, *182*, 613–614.
8. Fulton, B.; Goa, K. L. *Drugs* **1997**, *53*, 281–298.
9. Bever, K. A.; Perry, P. J. *Am. J. Health Syst. Pharm.* **1998**, *55*, 1003–1016.
10. Barnes, T. R.; McPhillips, M. A. *Int. Clin. Psychopharmacol.* **1998**, *13 Suppl 3*, S49–S57.
11. Raja, M. *Drug Saf.* **1998**, *19*, 57–72.
12. Umbricht, D.; Kane, J. M. *Schizophr. Bull.* **1996**, *22*, 475–483.
13. Jauss, M.; Schroder, J.; Pantel, J.; Bachmann, S.; Gerdsen, I.; Mundt, C. *Pharmacopsychiatry* **1998**, *31*, 146–148.
14. Owens, D. G. *Drugs* **1996**, *51*, 895–930.
15. Willner, P. *Int. Clin. Psychopharmacol.* **1997**, *12*, 297–308.
16. Sibley, D. R.; Monsma, F. J., Jr. *Trends Pharmacol. Sci.* **1992**, *13*, 61–69.
17. O'Dowd, B. *J. Neurochem.* **1993**, *60*, 804–816.
18. De Keyser, J. *Neurochem. Int.* **1993**, *22*, 83–93.
19. Seeman, P.; Tallerico, T. *Mol. Psych.* **1998**, *3*, 123–134.
20. Coffin, V. L.; Latranyi, M. B.; Chipkin, R. E. *J. Pharmacol. Exp. Ther.* **1989**, *249*, 769–774.
21. Carlsson, A.; Kehr, W.; Lindqvist, M. *J. Neural. Transm.* **1976**, *39*, 1–19.
22. Clark, D.; Hjorth, S.; Carlsson, A. *J. Neural Transm.* **1985**, *62*, 1–52.
23. Clark, D.; Hjorth, S.; Carlsson, A. *J. Neural Transm.* **1985**, *62*, 171–207.
24. Aghajanian, G. K.; Bunney, B. S. *Naunyn-Schmeideberg's Arch. Pharmacol.* **1977**, *297*, 1–7.
25. Sokoloff, P.; Giros, B.; Matres, M. P.; Andrieux, M.; Besancon, R.; Pilon, C.; Bouthnet, M. L.; Souil, E.; Schwartz, J. C. *Arzneim.-Forsch./Drug Res.* **1992**, *42*, 224–230.
26. Meller, E.; Bohmaker, K.; Goldstein, M.; Basham, D. A. *Eur. J. Pharmacol.* **1993**, *249*, R5–R6.
27. Wustrow, D. J.; Wise, L. D. *Curr. Pharm. Res.* **1997**, *3*, 391–404.
28. Waters, N.; Svensson, K.; Haadsma-Svensson, S. R.; Smith, M. W.; Carlson, A. *J. Neural Transm.* **1993**, *94*, 11–19.
29. Gifford, A. N.; Johnson, K. M. *Eur. J. Pharmacol.* **1993**, *237*, 169–175.
30. Svensson, K.; Carlsson, A.; Waters, N. *J. Neural Transm.* **1994**, *95*.
31. Meador-Woodruff, J. H.; Mansour, A.; Bunzow, J. R.; Van Tol, H. H.; Watson, S. J. Jr.; Civelli, O. *Proc. Natl. Acad. Sci. USA* **1989**, *86*, 7625–7628.
32. Helmreich, I.; Reimann, W.; Hertting, G.; Starke, K. *J. Neurosci.* **1982**, *7*, 1559–1566.
33. Carlsson, A. *J. Neural Transm.* **1983**, *57*, 309–315.
34. Meller, E.; Bohmaker, K.; Namba, Y.; Friedhoff, A. J.; Goldstein, M. *Mol. Pharmacol.* **1987**, *31*.
35. Meller, E.; Ena, A.; Goldstein, M. *Eur. J. Pharmacol.* **1988**, *155*, 151–154.
36. Carlsson, A. *Dopamine Autoreceptors and Schizophrenia*; Carlsson, A., Ed.; Cambridge University Press: Cambridge, 1988, pp. 1–10.

37. Creese, I.; Burt, D. R.; Snyder, S. H. *Science (Washington, DC)* **1976**, *192*, 481–483.
38. Pugsley, T. A.; Christofferson, C. L.; Corbin, A.; DeWald, H. A.; Demattos, S.; Meltzer, L. T.; Myers, S. L.; Shih, Y.-H.; Whetzel, S. Z.; Wiley, J. N.; Wise, L. D.; Heffner, T. G. *J. Pharmacol. Exp. Ther.* **1992**, *263*, 1147–1158.
39. Walters, J. R.; Roth, R. H. *Biochem. Pharmacol.* **1976**, *25*, 649–654.
40. Walters, J. R.; Roth, R. H. *Naunyn-Schmiedeberg's Arch. Pharmacol.* **1976**, *296*, 5–14.
41. Meltzer, L. T.; Christoffersen, C. L.; Corbin, A. E.; Ninteman, F. W.; Serpa, K. A.; Wiley, J. N.; Wise, L. D.; Heffner, T. G. *J. Pharm. Exp. Ther.* **1995**, *274*, 912–920.
42. Strombom, U. *Naunyn-Schmiedeberg's Arch. Pharmacol.* **1976**, *292*, 167–176.
43. Martin, G. E.; Bendesky, R. J. *J. Pharmacol. Exp. Ther.* **1984**, *229*, 706–711.
44. Svensson, L.; Ahlenius, S. *Eur. J. Pharmacol.* **1983**, *88*, 393–397.
45. Liebman, J.; Neale, R. *Psychopharmacol. (Berlin)* **1980**, *68*, 25–29.
46. Heffner, T. G.; Downs, D. A.; Meltzer, L. T.; Wiley, J. N.; Williams, A. E. *J. Pharmacol. Exp. Ther.* **1989**, *251*, 105–112.
47. Barany, S.; Ingvast, A.; Gunne, L. M. *Res. Commun. Chem. Pathol. Pharmacol.* **1979**, *25*, 269–279.
48. Lehmann, J.; Briley, M.; Langer, S. Z. *Eur. J. Pharmacol.* **1983**, *88*, 11–26.
49. Anden, N. E.; Nilsson, H.; Ros, E.; Thornstrom, U. *Acta Pharmacol. Toxicol. Copenh.* **1983**, *52*, 51–56.
50. Ericksson, E.; Svensson, K.; Clark, D. *Life Science (Washington, DC)* **1985**, *36*, 1819–1827.
51. Titus, R. D.; Kornfeld, E. C.; Jones, N. D.; Clemens, J. A.; Smalstig, E. B.; Fuller, R. W.; Hahn, R. A.; Hynes, M. D.; Mason, N. R.; Wong, D. T.; Foreman, M. M. *J. Med. Chem.* **1983**, *26*, 1112–1116.
52. Hansson, L. O.; Waters, N.; Holm, S.; Sonesson, C. *J. Med. Chem.* **1995**, *38*, 3121–3131.
53. Metman, L. V.; Sethy, V. H.; Roberts, J. R.; Bravi, D.; Hoff, J. I.; Mouradian, M. M.; Chase, T. N. *Mov. Disord.* **1994**, *9*, 577–581.
54. Tamminga, C. A.; Cascella, N. G.; Lahti, R. A.; Lindberg, M.; Carlsson, A. *J. Neural. Transm. Gen. Sect.* **1992**, *88*, 165–175.
55. VonVoigtlander, P. F.; Althaus, J. S.; Ochoa, M. C.; Neff, G. L. *Drug Dev. Res.* **1989**, *17*.
56. Piercey, M. F.; Broderick, P. A.; Hoffmann, W. E.; Vogelsang, G. D. *J. Pharmacol. Exp. Ther.* **1990**, *254*, 369–374.
57. Jaen, J., C. *Bioorg. Med. Chem. Lett.* **1991**, *1*, 189–192.
58. Benkert, O.; Muller Siecheneder, F.; Wetzel, H. *Eur. Neuropsychopharmacol.* **1995**, *5 Suppl.*, 43–53.
59. Coward, D. M.; Dixon, A. K.; Urwyler, S.; White, T. G.; Enz, A.; Karobath, M.; Shearman, G. *J. Pharmacol. Exp. Ther.* **1990**, *252*, 279–285.
60. Freeman, H.; McDermed, J. D.; Freeman, H.; McDermed, J. D., Ed.; Royal Society of Chemistry: London, 1982.
61. Wikstrom, H.; Sanchez, D.; Lindberg, P.; Hacksell, U.; Avidsson, L.-E.; Johansson, A. M.; Thorberg, S.-O.; Nilsson, J. L.; Svensson, K.; Hjorth, S.; Clark, D.; Carlsson, A. *J. Med. Chem.* **1984**, *27*, 1030–1036.

62. Bottcher, H.; Barnickel, G.; Hausberg, H. H.; Haase, A. F.; Seyfried, C. A.; Eiermann, V. *J. Med. Chem.* **1992**, *35*, 4020–4026.

63. Jaen, J. C.; Wise, L. D.; Heffner, T. G.; Pugsley, T. A.; Meltzer, L. T. *J. Med. Chem.* **1988**, *31*, 1621–1625.

64. Breese, G. R.; Traylor, T. D. *J. Pharmacol. Exp. Ther.* **1970**, *174*, 413–420.

65. Stricker, E. M.; Zigmond, M. J. *J. Comp. Physiol. Psychol.* **1974**, *86*, 973–994.

66. Jaen, J. C.; Wise, L. D.; Heffner, T. G.; Pugsley, T. A.; Meltzer, L. T. *J. Med. Chem.* **1991**, *34*, 248–256.

67. Coughenour, L. L.; McLean, J. R.; Parker, R. B. *Pharmacol. Biochem. Behav.* **1977**, *6*, 351–353.

68. Braun, A. R.; Chase, T. N. *Eur. J. Pharmacol.* **1986**, *131*, 301–306.

69. Arnt, J.; Hyttel, J.; Perregaard, J. *Eur. J. Pharmacol.* **1987**, *133*, 137–145.

70. Walters, J. R.; Bergstrom, D. A.; Carlson, J. H.; Chase, T. N.; Braun, A. R. *Science (Washington DC)* **1987**, *236*, 719–722.

71. Arnt, J.; Bogeso, K. P.; Hyttel, J.; Meier, E. *Pharmacol. Toxicol.* **1988**, *62*, 121–130.

72. Jaen, J. C.; Caprathe, B. W.; Wise, L. D.; Meltzer, L. T.; Pugsley, T. A.; Heffner, T. G. *Bioorg. Med. Chem. Lett.* **1993**, *3*, 639–644.

73. Wright, J. L.; Caprathe, B. W.; Downing, D. M.; Glase, S. A.; Heffner, T. G.; Jaen, J. C.; Johnson, S. J.; Kesten, S. R.; MacKenzie, R. G.; Meltzer, L. T.; et al. *J. Med. Chem.* **1994**, *37*, 3523–3533.

74. Pugsley, T. A.; Davis, M. D.; Akunne, H. C.; Cooke, L. W.; Whetzel, S. Z.; MacKenzie, R. G.; Shih, Y.-H.; Van Leeuwen, D. H.; Demattos, S. B.; Georgic, L. M.; Caprathe, B. W.; Wirght, J. C.; Jaen, J. C.; Wise, L. D.; Heffner, T. G. *J. Pharm. Exp. Ther.* **1995**, *274*, 898–911.

75. De Lean, A.; Stadel, J. M.; Lefkowitz, R. J. *J. Biol. Chem.* **1980**, *255*, 7108–7117.

76. Wright, J. L.; Downing, D. M.; Feng, M. R.; Hayes, R. N.; Heffner, T. G.; MacKenzie, R. G.; Meltzer, L. T.; Pugsley, T. A.; Wise, L. D. *J. Med. Chem.* **1995**, *38*, 5007–5014.

77. Wise, L. D.; Jaen, J. C.; Caprathe, B. W.; Smith, S. J.; Pugsley, T. A.; Heffner, T. G. *Soc. Neurosci. Abs.* **1991**, *17*, 689.

78. Wustrow, D. J.; Wise, L. D.; Cody, D. M.; MacKenzie, R. G.; Georgic, L. M.; Pugsley, T. A.; Heffner, T. G. *J. Med. Chem.* **1994**, *37*, 4251–4257.

79. O'Hara, C.; Uhland Smith, A.; KL, O. M.; Todd, R. D. *J. Pharmacol. Exp. Ther.* **1996**, *277*, 186–192.

80. Tang, L.; Todd, R. D.; KL, O. M. *J. Pharmacol. Exp. Ther.* **1994**, *270*, 475–479.

81. Belliotti, T. R.; Kesten, S. R.; Rubin, J. R.; Wustrow, D. J.; Georgic, L. M.; Zoski, K. T.; Akunne, H. C.; Wise, L. D. *Bioorg. Med. Chem. Lett.* **1997**, *7*, 2403–2408.

82. Wustrow, D., J.; Smith, W. J. I.; Corbin, A. E.; Davis, M. D.; Georgic, L. M.; Pugsley, T. A.; Whetzel, S. Z.; Heffner, T. G.; Wise, L. D. *J. Med. Chem.* **1997**, *40*, 250–259.

83. Glase, S. A.; Akunne, H. C.; Heffner, T. G.; Jaen, J. C.; MacKenzie, R. G.; Meltzer, L. T.; Pugsley, T. A.; Smith, S. J.; Wise, L. D. *J. Med. Chem.* **1996**, *39*, 3179–3187.

84. Wetzel, H.; Hillert, A.; Grunder, G.; Benkert, O. Am. *J. Psychiatry* **1994**, *151*, 1499–1502.

85. Wustrow, D.; Belliotti, T.; Glase, S.; Kesten, S. R.; Johnson, D.; Colbry, N.; Rubin, R.; Blackburn, A.; Akunne, H.; Corbin, A.; Davis, M. D.; Georgic, L.; Whetzel, S.; Zoski, K.; Heffner, T.; Pugsley, T.; Wise, L. *J. Med. Chem.* **1998**, *41*, 760–771.

86. McMillen, B. A.; Scott, S. M.; Davanzo, E. A. *J. Pharm. Pharmacol.* **1988**, *40*, 885–887.

87. Invernizzi, R. W.; Cervo, L.; Samanin, R. *Neuropharmacology* **1988**, *27*, 515–518.

88. Neal Beliveau, B. S.; Joyce, J. N.; Lucki, I. *J. Pharmacol. Exp. Ther.* **1993**, *265*, 207–217.

89. Wadenberg, M. L.; Cortizo, L.; Ahlenius, S. *Pharmacol. Biochem. Behav.* **1994**, *47*, 509–513.

NONPEPTIDE INHIBITORS OF HIV PROTEASE

Susan Hagen, J.V.N. Vara Prasad, and
Bradley D. Tait

Advances in Medicinal Chemistry
Volume 5, pages 159–195.
Copyright © 2000 by JAI Press Inc.
All rights of reproduction in any form reserved.
ISBN: 0-7623-0593-2

ABSTRACT

Mass screening of our compound collection, utilizing the Amersham SPA technology, afforded pyrone and coumarin non-peptide templates as initial lead structures. X-ray cocrystallization and structure-based design were utilized to assist in the design of more potent inhibitors. These efforts resulted in the design of the 5,6-dihydropyrones, which afforded a more flexible template from which to fill the internal pockets of the enzyme. Optimization of the dihydropyrone series afforded a potent antiviral agent, PD 178390 (EC_{50} = 0.20 μM, TD_{50} = >100 μM). PD 178390 retained antiviral potency in the presence of serum proteins with a modest three- to fivefold drop in antiviral activity in the presence of 40% human serum. The antiviral activity, in PBMCs, was unchanged against clinical strains of resistant HIV virus. In addition, PD 178390 showed excellent bioavailability in mice, rats, and dogs as well as a low level of P450 inhibition in microsomal assays. This combination of good antiviral efficacy, good pharmacokinetics, and low P450 inhibition make PD 178390 a promising agent for the treatment of HIV infection.

I. INTRODUCTION

Since the identification of the human immunodeficiency virus (HIV) as the causative agent of AIDS, the pharmaceutical and research communities have focused enormous energy and resources on the development of anti-retroviral chemotherapies. The reason behind such intense focus is obvious: it has been estimated that by the year 2000 more than 30 million individuals will be infected with HIV.[1] Therefore, the search for new targets for therapeutic intervention continues unabated as studies reveal further details of the HIV life cycle. One target of particular interest is HIV protease, an essential viral enzyme required for the cleavage of viral polypeptides into functional enzymes.[2] Inhibition of the HIV-1 protease results in the production of virions that are incapable of maturation and infection. Not only is the HIV protease essential to viral replication, but it has been repeatedly isolated and crystallized as inhibitor–enzyme complexes.[3] Such complexes have provided a wealth of information invaluable in the design of highly potent, highly specific protease inhibitors. Indeed, the development and use of protease inhibitors has revolutionized AIDS care.[4]

However, the success story of the protease inhibitors has been muted by the emergence of several disturbing trends. Many of the currently marketed inhibitors suffer from low bioavailability, substantial protein binding, and short half-lives.[5] These attributes result in drugs that are expensive and inconvenient to take.[6,2d] In addition, significant side effects and drug interactions restrict the usefulness of these agents in real-world scenarios.[6,7] Most ominous is the development of HIV strains that are resistant to almost all current therapies.[8] Therefore, the need for novel protease inhibitors which are not cross-resistant to the current generation of agents remains.

In many cases, the focus of new research has shifted to synthetic non-peptidic molecules of low molecular weight. Towards this end, we at Parke-Davis initiated a screening strategy to identify a nonpeptide hit suitable for further optimization. The implementation of this strategy and the optimization of the resulting lead are the focus of this review article.

II. FINDING AND EVALUATING AN INITIAL LEAD

We screened our compound collection (consisting of approximately 150,000 chemical entities) with a high throughput SPA (Scintillation Proximity Assay) developed by Amersham to find an initial lead for further optimization. The assay was done in 96 well plates utilizing a β-scintillation counter to monitor the reaction. We initially screened the compounds as solutions of 10 compounds at ~100 μM each. Any well with greater than 50% inhibition was deconvoluted into a single compound per well at 50 μM, 30 μM, and 10 μM concentrations and assayed using the SPA technology. Compounds that had an IC_{50} of less than 50 μM were then run in an HPLC assay.[9]

After our analysis we ended up with about 15 series which were clustered on the basis of their structures.[10] We prioritized the series based upon the following criteria:

1. non-peptidic structures;
2. selectivity for HIV protease (leads should not be hits in multiple mass screens);
3. purity and integrity of the sample;
4. reasonable modes of binding to HIV protease (hits were docked into the protease active site); and
5. competitive inhibition (as estimated by kinetic analysis).

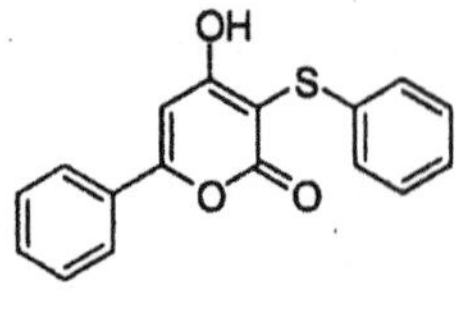

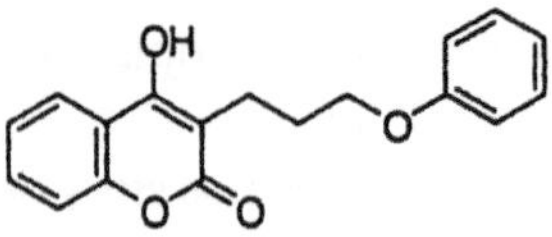

PD 107067   IC$_{50}$ = 3.1 μM   **PD 099560**   IC$_{50}$ = 2.3 μM

Figure 1. Initial mass screen hits.

Additional background information such as other biological activity, toxicity, patent status, and physical properties was also obtained to better understand what data was available on the series in question. Finally, additional congeners were pulled from our compound collection and tested in the HPLC assay to get an initial read on the structure–activity relationships (SAR).

After a full analysis of the hits we focused on the coumarin (PD 099560 IC$_{50}$ = 2.3 μM) and pyrone (PD 107067, IC$_{50}$ = 3.1 μM) as viable compounds for further elaboration (Figure 1). Both Parke-Davis[9–31] and Pharmacia-Upjohn[32–46] have reported extensively on their work. We will review our effort starting with the initial leads and optimization of the leads into a potent biologically active compound, PD 178390.

III. COUMARINS (4-HYDROXYBENZOPYRAN-2-ONES)

Both Parke-Davis and Upjohn researchers have identified warfarin and its analogue as weak inhibitors of HIV protease.[9,32] Furthermore, Bourinbaiar et al.[47,48] have also reported that warfarin possessed an antiviral effect on HIV replication and spread, but it was unclear if this antiviral

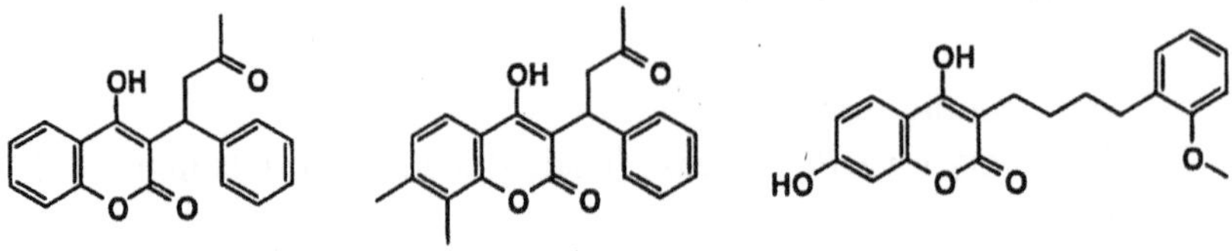

1 (Warfarin)   **2**   **3**

IC$_{50}$ = 30 μM   IC$_{50}$ = 1.9 μM   IC$_{50}$ = 0.52 μM

Figure 2. Coumarin inhibitors.

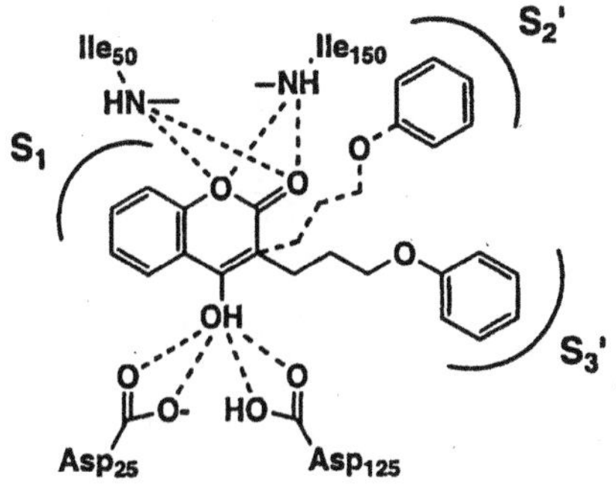

Figure 3. X-ray crystal structure of PD 099560 bound to HIV PR.

activity was due to inhibition of HIV-1 protease. Among the various warfarin analogues tested Tummino et al. found that inhibitor **2** (Figure 2) was the most potent analogue.

At Parke-Davis, mass screening also identified another coumarin analogue (PD 099560, Figure 1), as a competitive inhibitor of the enzyme.[13,25] X-ray crystallographic structure of PD 0099560 bound to HIV PR (Figure 3) revealed two binding modes. In both modes, the 4-hydroxycoumarin ring of the inhibitor displaced both water molecules (the catalytic water as well as water-301) and the fused phenyl ring was oriented in the S_1 site. Active site interactions between the aspartates and the 4-OH were similar to those observed in the pyrone X-ray. In one binding mode, the flexible side chain extended to the S_3' region towards Arg 108 while in the other mode, the side chain folded back down to the S_2' region. With this data in hand, several analogues were prepared to test the binding interactions and improve the overall binding affinity. Among them the best inhibitor **3** was found to possess an IC_{50} of 0.52 μM (Figure 2). Unfortunately, extensive synthetic modification and substitution of the coumarin ring system did not lead to significant increases in enzyme potency for this series.

IV. PYRONES (4-HYDROXYPYRAN-2-ONE)

Though 4-hydroxypyran-2-one derivatives were first reported in 1993 as anti-HIV agents[47,48] neither the mechanism of action nor the mode of interactions with the protein was known until the recent reports from Parke-Davis and Upjohn. Currently, various derivatives of the 4-hy-droxypyran-2-one (e.g. 4-hydroxybenzopyran-2-ones (coumarins), sub-

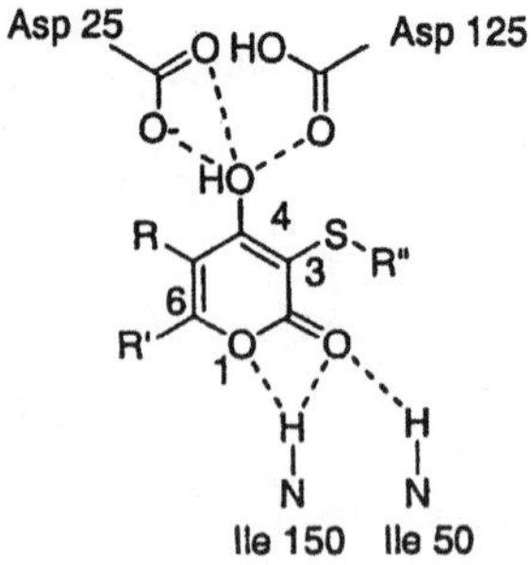

Figure 4. Coumarin and pyrone core interactions.

stituted 4-hydroxypyran-2-ones, 4-hydroxycyclooctapyran-2-ones and 4-hydroxy-5,6-dihydropyran-2-ones)[9–46] have been reported to be potent inhibitors of HIV protease. Both kinetic information and X-ray crystallographic data of these pyran-2-ones (Figure 4) reveal that these compounds act on HIV protease via a common mode of binding with the enzyme in the active site. In particular, the 4-hydroxypyran-2-one replaces the water molecules found in the active site of the enzyme, while the 4-hydroxyl group forms hydrogen bonds with the catalytic aspartates (Asp25 and Asp125). The lactone moiety forms hydrogen bonds with the flaps Iles (Ile50 and Ile150) by replacing water-301, a unique water molecule found in all the X-ray crystallographic structures of peptide-derived inhibitors binding to HIV protease.[3] It was theorized that various groups could be appended to this pyrone core, groups that could interact with the binding pockets of the protease enzyme. Such interactions would thereby result in improvement in inhibitory activity against HIV protease.

V. ELABORATION OF THE
6-ARYL-4-HYDROXYPYRAN-2-ONES

The 4-hydroxy-6-phenyl-3-(phenylthio)pyran-2-one (PD 107067, Figure 1) fulfilled all our initial criteria for a viable series suitable for further elaboration. By chemical analogy with the known peptidomimetic hydroxyethylsulfide (HES),[48] PD 107067 was viewed as a conformationally restricted P_1-P_1' peptidomimetic (Figure 5).[14]

Initially, modifications at the C-3 position were undertaken to explore the optimal chain length at the S–Ph region. Extending the S–Ph moiety to SCH_2Ph and SCH_2CH_2Ph produced compounds **5** and **6**, respectively

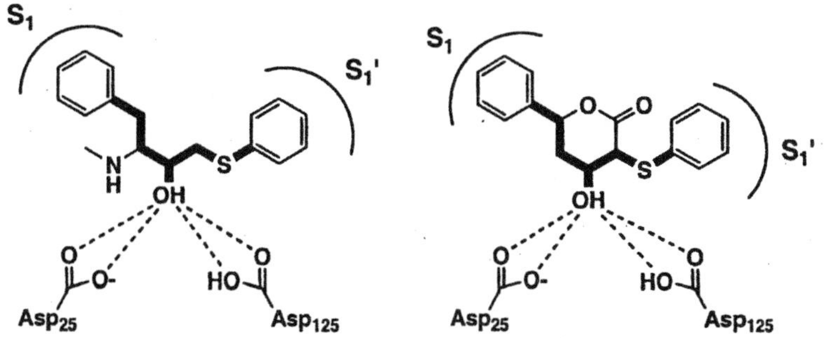

Figure 5. Comparison of hydroxyethyl sulfide isostere with PD 107067.

(Figure 6), both of which displayed a two- to threefold enhancement in enzyme activity. An X-ray crystal structure of **5** (see Figure 7) bound to HIV protease showed a tight binding interaction between the SCH_2Ph group and the S_1' pocket. As expected, the core interactions between the pyrone nucleus and the active site remained unchanged.

Substitution of the phenyl group in the SCH_2Ph moiety was evaluated, in the hopes that the appropriate substituent might fill an additional

The S-Ph at C-3:

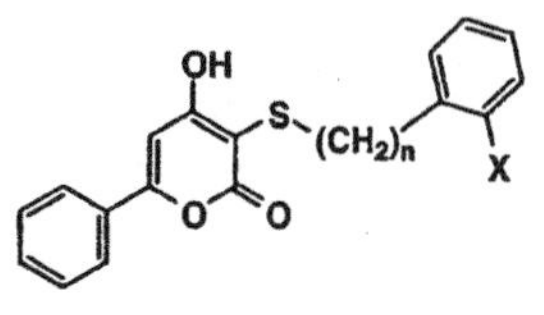

4, n = 0, X = H, IC_{50} = 3.1 μM
5, n = 1, X = H, IC_{50} = 1.7 μM
6, n = 2, X = H, IC_{50} = 1.3 μM
7, n = 1, X = $CO_2(i$-Propyl), IC_{50} = 0.034 μM

The Phenyl at C-6:

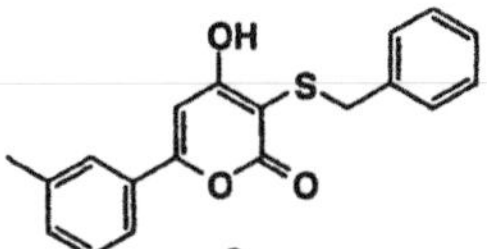

8

IC_{50} = 0.26 μM

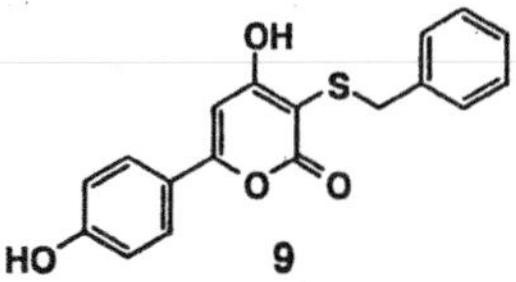

9

IC_{50} = 0.52 μM

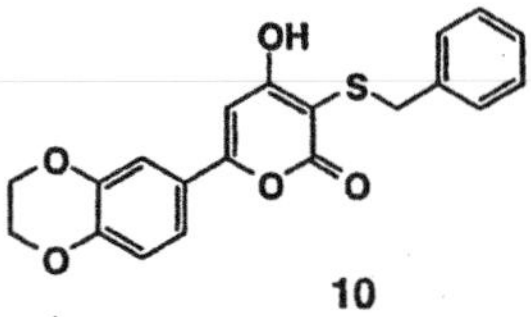

10

IC_{50} = 0.49 μM

Figure 6. SAR of Pyrones.

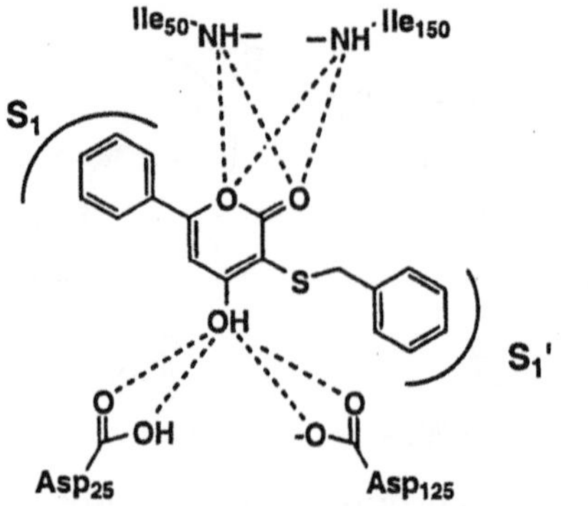

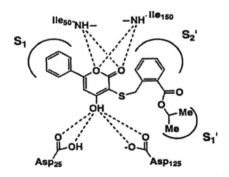

Figure 7. X-ray crystal structure of compound **5** bound to HIV PR.

binding pocket. Systematic modification of this phenyl group led to a series of benzyl esters with increased enzymatic activity.[26] The most potent derivative in this subclass, compound **7** contained an isopropyl ester adjacent to the sulfur linkage and is shown bound to the protease enzyme in Figure 8. Surprisingly, this X-ray crystal structure revealed a change in binding mode: for compound **7** the SCH$_2$Ph group is oriented in the S$_2'$ pocket pocket while the CO$_2$(isopropyl) group fills the S$_1'$ pocket. This observation differs decidedly from the X-ray crystal structure for compound **5** discussed above, in which the unsubstituted SCH$_2$Ph group occupies the S$_1'$ pocket pocket.

Attention then shifted to the 6-position of the pyrone molecule. Variation of the substituents on the phenyl ring at C-6 resulted in a series of derivatives with improved enzyme activity. More specifically, the

Figure 8. X-ray crystal structure of compound **7** bound to HIV PR.

meta-methyl analogue **8**, the para-hydroxy analogue **9**, and the 3,4-benzodioxyl analogue **10** proved to be the most potent entities in this class (see Figure 6). Apparently, a small hydrophobic group at the meta position and a hydrophilic group para are well tolerated.

Therefore, appropriate substitution at C-3 and C-6 resulted in inhibitors binding to two (and sometimes 3) pockets of the protease enzyme, and a concomitant increase in activity was observed. Occupation of other binding pockets would conceivably lead to further significant enhancements in potency. Toward this end, molecular modeling and X-ray analysis suggested that branching at C-3 might achieve simultaneous binding at both S_1' and S_2'. For this reason, a series of (4-hydroxy-6-phenyl-2-oxo-2*H*-pyran-3-yl)thiomethanes was synthesized and their binding affinities were measured.[15] The results are summarized in Table 1.

Both S–aryl and S–aliphatic groups, having different steric and hydrophobic properties, were introduced to optimize the molecular recognition and the binding affinity. In the S–aryl series **11–18** (phenyl and benzyl), the cylopropylmethyl group was the most effective substitution at the R_2 position, whereas phenyl and benzyl groups were well accommodated at the SR_1 position. Overall, the S–aliphatic series, (**19–25**, Table 2) showed better binding affinity relative to the S–aryl

Table 1. 4-Hydroxypyran-2-ones Having S-Aryl Functionalization at C-3

Compound	R_1	R_2	IC_{50} (µM)
11	Ph	H	84.3
12	Ph	Ph	0.78
13	Ph	cyclohexyl	2.44
14	Ph	isobutyl	0.41
15	Ph	isopentyl	0.39
16	Benzyl	Ph	0.48
17	Benzyl	isobutyl	0.26
18	Benzyl	CH_2cPr	0.084

Table 2. 4-Hydroxypyran-2-ones Having S-Aliphatic Functionalization at C-3

Compound	R_1	R_2	IC_{50} (μM)
19	Cyclohexyl	Ph	0.48
20	Cyclohexyl	isobutyl	0.32
21	Cyclohexyl	CH_2cPr	0.15
22	Cyclohexyl	neopentyl	0.30
23	Cyclopentyl	cyclopentyl	0.22
24	Cyplopentyl	isobutyl	0.058
25	Cyclopentyl	CH_2cPr	0.069

26			0.037

series. Unlike the S–aryl series, both branched and cyclic aliphatic substituents impart potency at the R_2 position. Replacement of the 6-phenyl group with a 6-(benzodioxyl) group (**26**) resulted in a similar or slight improvement in binding affinity. Kinetic analysis of inhibitors **24** and **26** showed that they are competitive inhibitors with K_i values 33 and 27 nM, respectively.

The X-ray crystal structure of **24** bound to HIV-1 PR (Figure 9) showed a unique mode of binding which was not observed previously with other inhibitors. In this case, the lactone carbonyl of the pyran-2-one ring formed a direct hydrogen bond with NH of Ile50 only. This result contrasts to the X-ray crystal structure of structurally related inhibitor **5** (Figure 7), in which the lactone is positioned more symmetrically to form hydrogen bonds with both Ile50 and Ile150. The enol moiety of **24** forms hydrogen bonds with Asp125 and interacts indirectly with Asp25 via a bridging water molecule. This bridging water has not been observed in any of the HIV PR crystal structures

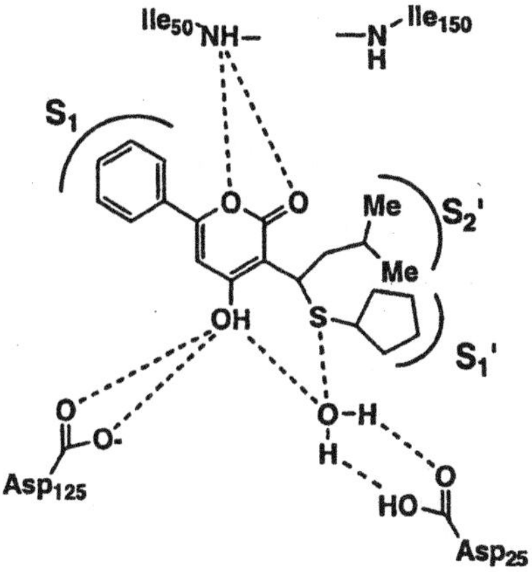

Figure 9. X-ray crystal structure of compound **24** bound to HIV PR.

reported to date. The water molecule is 3.4 Å away from the sulfur atom present in the inhibitor. The S-cyclopentyl and isobutyl occupy the S_1' and S_2' pockets, respectively, whereas the 6-phenyl group straddles the S_1 and S_3 pockets of the enzyme.

The strategies used to fill the S_1' and S_2' pockets involved branching at C-3 position and therefore resulted in the formation of a new chiral center. Our efforts to eliminate chirality and simultaneously improve binding affinity resulted in an interesting series of pyran-2-one analogues shown in Table 3.[20] This strategy (Figure 10) evolved from a literature report that α-alkylbenzamides can be incorporated into HIV PR inhibitors as P_1' proline mimics.[51] Adapting this idea to our pyran-2-one series and replacing the amide portion with simple alkyl re-

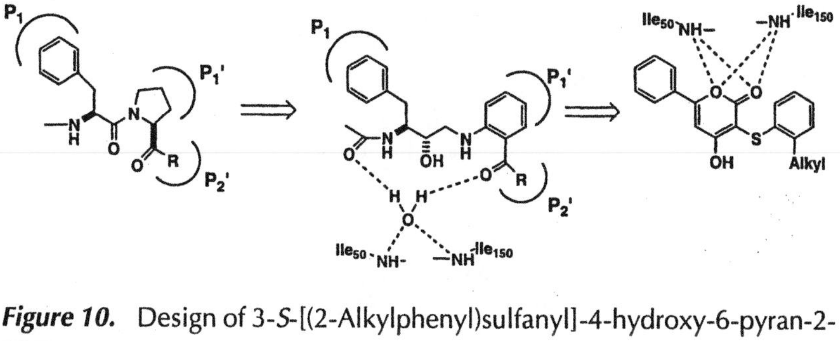

Figure 10. Design of 3-S-[(2-Alkylphenyl)sulfanyl]-4-hydroxy-6-pyran-2-ones.

Table 3. 4-Hydroxypyran-2-ones: Substitution on the 3-SPh Ring

Compound	R	IC_{50} (μM)
27	H	3.0
28	Me	0.42
29	Et	0.17
30	*n*-propyl	0.090
31	isopropyl	0.037
32	*sec*-butyl	0.10
33	*tert*-butyl	0.017
34	CH_2cyclopropyl	0.10

sulted in a 3-*S*-(2-alkylphenyl)group as the P_1'/P_2' ligand (Figure 10). Optimization of the size of the alkyl group lead to an inhibitor, **33**, possessing a K_i of 3 nM (Table 3).

The X-ray crystal structure of **31** bound to HIV-1 PR revealed the typical hydrogen-bonding interactions of central pyran-2-one ring as described previously. In contrast to our initial hypothesis, but in line with the molecular modeling prediction, the isopropyl group present on the 3-*S*-phenyl group occupied the S_1' pocket, whereas the 3-*S*-phenyl group partially filled the S_2' pocket (Figure 11). This analogue occupies only three pockets of the enzyme, yet possesses very high

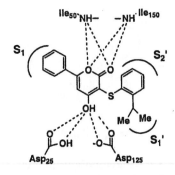

Figure 11. X-ray crystal structure of compound **31** bound to HIV PR.

Table 4. 4-Hydroxypyran-2-one: Tethers on the *para* Position of the 6-Phenyl

R	Compound	IC_{50} (μM)	Compound	IC_{50} (μM)
OCH$_2$COOH	**35**	0.16	**36**	0.019
OCH$_2$Ph	**37**	0.33	**38**	0.28
OCH$_2$(3-pyridine)	**39**	1.5	**40**	0.045

binding affinity to HIV PR. To further optimize activity, a Topliss operationalscheme[50] was used for substitution on the phenyl ring the 6-position, with modest results.[23]

Since the S$_1$ and S$_3$ pockets are topographically contiguous, it was envisioned that a group appended to the C-6 phenyl group (which occupies the S$_1$ pocket) could reach into the S$_3$ pocket via a judiciously chosen tether. In general, the tethering approach[51] might fill either the S$_3$ pocket or protrude through the active site openings to the solvent; these tethers could also be useful in altering the physical properties of the inhibitor.[52] Thus pyran-2-one analogues with various tethers at the para and meta positions of the 6-phenyl ring were prepared in both the 3-SCH$_2$CH$_2$Ph series and 3-SPh(2-*i*Pr) series.[27,28] The results are described in Table 4. Initially a tether containing OCH$_2$COOH at the para position of the 6-phenyl ring was designed to interact with Arg-8 present in the S$_3$ pocket of the enzyme. The inhibitors **35** and **36**, obtained from 3-SCH$_2$CH$_2$Ph series and 3-SPh(2-*i*Pr) series, respectively, showed a five- and twofold enhancement in binding affinities compared to the parent compounds. Since aromatic hydrophobic groups are known to occupy S$_3$ pocket of the enzyme, various hydrophobic groups were also used.[53] It was found that a tether from the para position was optimal for binding to HIV PR (Table 4).

VI. TRICYCLIC-4-HYDROXYPYRAN-2-ONES

The tricyclic 4-hydroxypyran-2-ones (Figure 12) were synthesized as an intermediate structure between the coumarins and 6-aryl-4-hy-

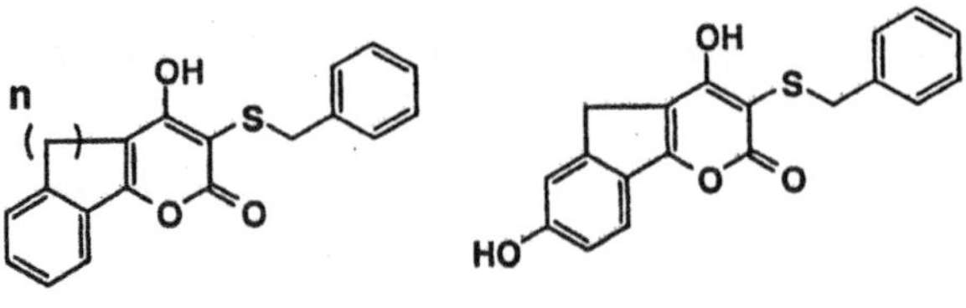

41, $IC_{50} = 1$ μM

Figure 12. Tricyclic 4-hydroxypyran-2-ones.

droxypyran-2-ones. The analogues containing SCH_2Ph or SCH_2CH_2Ph moiety at the 3-position did not show improved binding affinity when compared to 6-aryl-pyran-2-one series. The best compound in the series, **41**, possesses an IC_{50} of 1 μM against HIV PR.

VII. 5,6-DIHYDROPYRONES

The coumarin and pyrone ring systems are rigid and planar and thus groups attached to the ring systems are locked into position relative to each other and relative to the core binding interactions. We hypothesized that the 5,6-dihydropyrones[10,21,24,30,31] would offer a more flexible template that would allow the substituents to make modest conformational adjustments without interfering with the core binding.

Molecular modeling of the dihydropyrone template was carried out using the Sybyl software program and a preliminary X-ray structure of HIV protease from the PD 099560 complex. Due to the structural similarities among the 4-hydroxybenzopyran-2-one, the pyrone, and the dihydropyrone templates, we believed that a similar core-binding mode would occur in all the templates. We therefore turned our attention to exploring ways of filling the S_2 and S_1 pockets. Our model of the 6-phenylpyrones such as PD 107067 (Figure 1) indicated that the 6-position substituent filled S_1. It was apparent from analysis of a coumarin X-ray and the subsequent dihydropyrone model that S_2 could be reached from the sp^3 carbon atom at the 6-position of the dihydropyrone by appending an appropriate substituent (R in Figure 13).

Inhibitor **47b** (Table 5) was modeled in the active site with the S configuration. The phenyl ring was placed in the equatorial position and bound in S_1 and the isopentyl group was in the axial orientation and extended to S_2 (Figure 14). Figure 14 shows a model of **47b** overlaid with the Abbott HIV protease inhibitor (A-74704). The Abbott structure

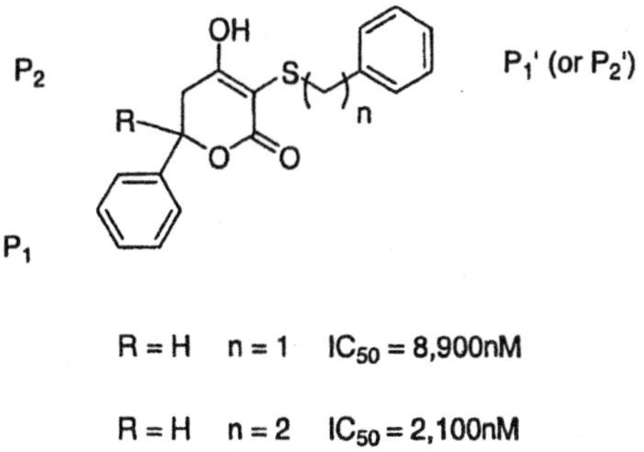
P2
P1' (or P2')
P1
R = H n = 1 IC50 = 8,900nM
R = H n = 2 IC50 = 2,100nM

Figure 13. 4-Hydroxy-5,6-dihydropyrones: P_1–P_1' (or P_2').

Table 5. Hydrophobic Groups Designed to Fill S_2: P_2P_1–P_1' (or P_2')

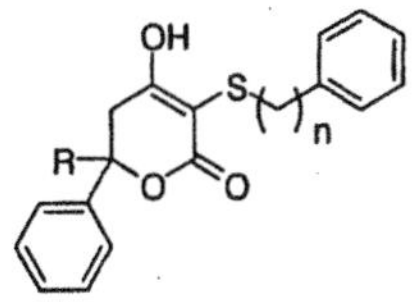

		IC_{50} (nM)	
		a	*b*
Compound[a]	*R*	*n = 1*	*n = 2*
41a,b	H	8,900	2,100
42a,b	$(CH_2)_2Me$	290	5100
43a,b	$(CH_2)_3Me$	270	400
44a,b	$(CH_2)_4Me$	124	84
45a,b	$(CH_2)_5Me$	180	112
46a,b	CH_2CHMe_2	440	1,200
47a,b	$(CH_2)_2CHMe_2$	98	96
48a,b	$(CH_2)_3CHMe_2$	183	155
49a,b	CH_2-*c*Pentyl	88	510
50a	CH_2-*c*Hexyl	137	
51a,b	Ph	262	152
52a,b	$(CH_2)_2Ph$	60	51

Note: [a]Where **a** is n = 1 and **b** is n = 2.

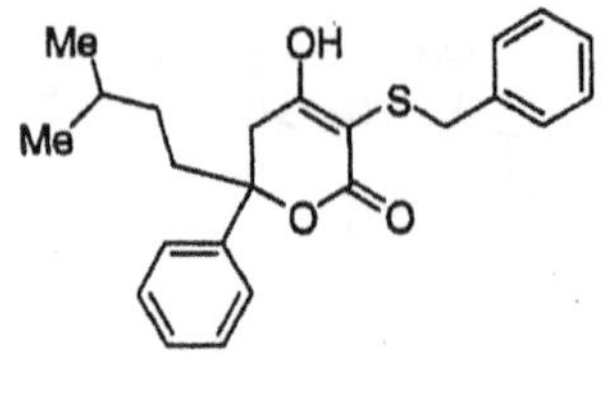

47b

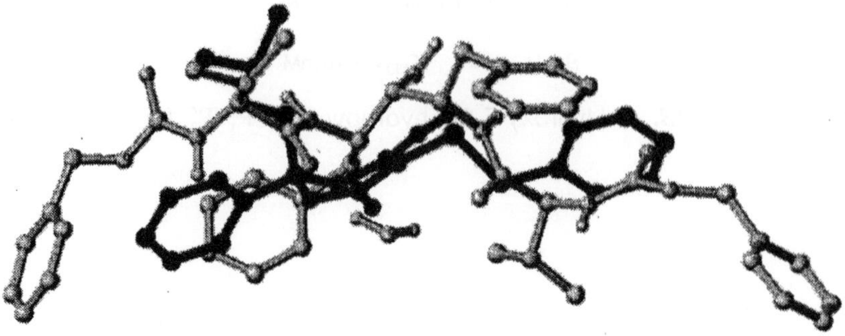

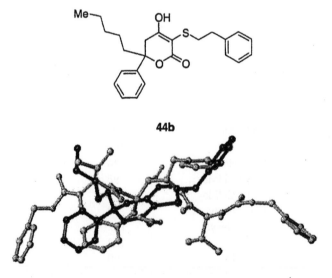

Figure 14. Model of **47b** overlayed with A-74704.

44b

Figure 15. X-ray of **44b** overlayed with A-74704.

assists in the visualization of the approximate position of P_1 and P_2 but does not define the surface area of the pockets. Our model indicated that from the 6-position we needed a 2-atom spacer followed by the substituent to fill S_2. The model also predicted that a phenyl at the 6-position was a reasonable substituent to fill S_1. Therefore, we held the phenyl constant and varied the R group to reach for S_2 (Table 5).

Our initial analysis led to the preparation of straight-chain alkyl groups at C-6 (see Table 5, **42–45**). Due to their flexibility, these straight-chain alkyl groups could readily adjust to the surface of the enzyme to maximize hydrophobic interactions, albeit with an entropic price. In any event, the potency increased as the chain was lengthened from hydrogen ($IC_{50} = 2100$ nM, **41b**) to *n*-pentyl ($IC_{50} = 84$ nM, **44b**).

Branched alkyl groups designed to mimic the P_2 and P_2' hydrophobic amino acids, valine and leucine (**46–48**, Table 5) were then investigated. Modeling predicted that isopentyl (**47**) and isohexyl (**48**) substituents would most closely mimic valine and leucine, respectively. These branched derivatives showed an increase in potency over the unsubstituted parent ($IC_{50} = 2100$ nM, **41b**) with the most potent compound being the isopentyl substituent ($IC_{50} = 96$ nM, **47b**), the valine mimic.

Workers at Merck have reported that phenyl glycine is an effective P_2' replacement for valine in peptidic HIV protease inhibitors. In a logical extension of their work, we prepared the phenethyl derivatives (**52**) as phenyl glycine mimics. Compound **52b** showed excellent activity versus the enzyme with an IC_{50} of 51 nM.

Our model indicated, and the activity confirmed, that the isobutyl group (**46**) was not reaching far enough to fill S_2. To more effectively fill the S_2 pocket, the two methyls of the isobutyl group were expanded and cyclized into five-membered ring (**49**) and six-membered ring (**50a**). In fact, compounds **49a** and **50a** were better than compound **46a** by more than threefold, thus suggesting enhanced interactions with the enzyme.

An X-ray cocrystal structure of racemic **44b** bound into HIV-1 protease was obtained (Figure 15). The general core interactions were consistent with those found in the coumarin X-ray and the dihydropyrone model. The presence of electron density in both S_1' and S_2' indicates a mixed population for the thiophenethyl group between the two pockets. It was hoped that only the more potent enantiomer would be bound in the active site but the X-ray indicated that both enantiomers of **44b** were present. Therefore, an accurate determination of the preference for the phenyl and *n*-pentyl groups to reside in S_1 or S_2 could not be obtained from the electron density. What is clear is that the substituents (phenyl and *n*-pentyl) fill both S_2 and S_1, thus increasing the potency of these

262 nM range (**51**), although not as potent as the racemic 6-phenyl-6-*n*-pentyl parent (**44**).

We next turned our attention to filling S_2 with side chains that mimic the more polar amino acids—asparagine, glutamine, and glutamic acid (Table 6). Asparagine has been used at S_2 in peptidic inhibitors such as Ro-31,8959 with excellent success. Modeling suggested that the optimal side chains should be $-(CH_2)_3-CONH_2$, $-(CH_2)_4-CONH_2$, and $-(CH_2)_4-COOH$, respectively. The most potent compound was carboxylic acid **55** with an IC_{50} of 5 nM. The potency decreased by almost 2 orders of magnitude when the carboxylic acid (**55**) was replaced with a primary amide (**57**).

Work in the pyrone series delineated the beneficial effect of substituting ortho to the sulfur linkage on the S–phenyl with an isopropyl moiety. Modeling of the ortho isopropyl functional group with the 5,6-dihyropyrone template indicated that it probably filled the S_1' pocket This observation led to modifications of the initial dihydropyrone lead, resulting in agents with improved enzyme potency (Table 7).

To address the question of steric requirements of S_1' (Table 7) a number of derivatives containing ortho-alkyl groups were prepared. Reducing the isopropyl group (**60**) to a methyl decreased the activity by 5- to 10-fold (**59**); increasing the size to *sec*-butyl (**61**) or cyclohexyl (**62**) instead of isopropyl (**60**) provided equally potent compounds. Relative to isopropyl, a *tert*-butyl moiety engendered a four- to fivefold increase in activity (IC_{50} of 3 nM, **63**).

Table 6. Polar Groups Designed to Fill S_2: $P_2P_1-P_1'$ (or P_2')

Compound	R.	IC_{50} (nM)
53	$(CH_2)_2CO_2H$	1,200
54	$(CH_2)_3CO_2H$	270
55	$(CH_2)_4CO_2H$	5
56	$(CH_2)_3CONH_2$	1,300
57	$(CH_2)_4CONH_2$	338

Table 7. Optimization of $P_1'P_2'$ Substituent: $P_2P_1-P_1'P_2'$

Compound	R'	R''	IC_{50} (nM)
58	H	H	130
59	Me	H	73
60	$CHMe_2$	H	14
61	CHMeEt	H	10
62	cHexyl	H	32
63	CMe_3	H	3
64	$CHMe_2$	Me	7
65	$CHMe_2$	$CHMe_2$	14
66	CMe_3	Me	10
67	CMe_3	$CHMe_2$	47
68(S)	$CHMe_2$	Me	6
69(R)	$CHMe_2$	Me	130

Assuming that the ortho substituent effectively satisfied S_1', attention turned to more optimally filling the S_2' pocket from the appropriate position of the phenyl ring (**64–67**). To determine that optimal substitution, a series of analogues was synthesized in which the ortho substituent was held constant as an isopropyl or *tert*-butyl group while varying the position para to the alkyl moiety. The methyl isomers display an enhancement in potency (**64**). The size of the methyl group appeared to be somewhat optimal, since the corresponding isopropyl compounds (**65, 67**) all displayed a loss in activity versus the methyl derivatives (**64, 66**).

In general, the inhibitory activity data indicates that a 2-*tert*-butyl/2-isopropyl group and 5-methyl moiety on the phenyl ring effectively fill the S_1' and S_2' pockets, respectively. To verify this model, an X-ray crystallographic study of racemic **64** with the protein was undertaken; unfortunately, from the electron density it was not possible to determine either the preferred enantiomer for binding or whether both enantiomers are bound. To overcome this problem, the (*R*)- and (*S*)-enantiomers were prepared and tested (**68, 69**); the (*S*)-enantiomer (**68**) proved to be 20-fold more potent the corresponding (*R*)-enantiomer (**69**).

The X-ray crystallographic structure of **68** complexed with the HIV protease enzyme was obtained (Figure 16). As expected the interactions

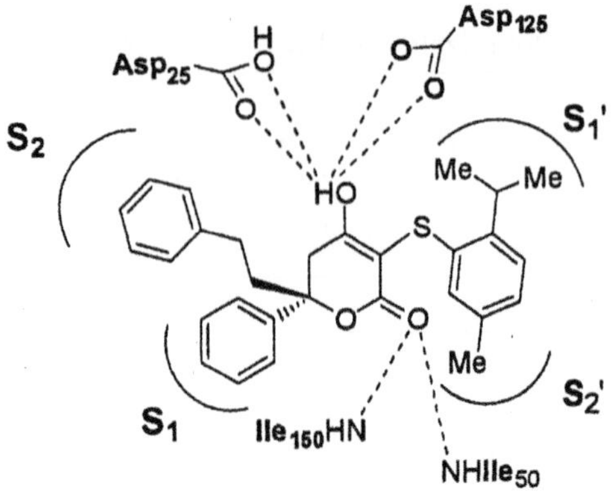

Figure 16. X-ray of compound **68**.

between the dihydropyrone ring and the protease active site were consistent with the previous X-rays. Also as predicted, the 3-position substituent occupies the S_1' site via the isopropyl group and the S_2' site site via the 5-methyl substituent. The (S)-enantiomer binds the phenyl group in the S_1 site and orients the phenethyl chain in the S_2 pocket. The (S)-enantiomer, predicted to be the more potent inhibitor, displays a binding affinity of 6 nM, whereas the (R)-enantiomer has an $IC_{50} = 130$ nM. This is in contrast to reports by Upjohn where the 6-phenethyl occupies S_1/S_3 and a 6-n-propyl occupies S_2. Although the Upjohn and Parke-Davis groups were both working on the dihydropyrone template, the preferred chirality of the substituents at the 6-position is apparently reversed.

A. Cellular Activity

In spite of obtaining a 1000 fold improvement in IC_{50}, the best compounds did not show a significant therapeutic index (TD_{50}/EC_{50} of > 10-fold) in cellular assays and therefore did not warrant claims of antiviral activity. The lack of cellular activity despite IC_{50}'s in the 3–10 nM range necessitated a close examination of possible explanations.

One area of concern was the pK_a of the 4-hydroxyl group ($pK_a = 4.5$–6.5) which is more acidic than a normal alcohol. We believe the protonated form of the dihydropyrone is the bound form of the inhibitor in the active site, and therefore, ionization decreases the quantity of the active protonated form. Thus, the pH of the assay would be expected to have an effect on inhibitory activity. To test this theory we took the best compounds and ran the assay at a pH of 6.2 rather than at a pH of 4.7. We found the dihydropyrones did lose potency at the higher pH (Table 8). Although all the compounds decreased in activity, the 2-*tert*-butyl-5-methyl compound (**66**) retained more of its inhibitory efficacy than did the other dihydropyrone derivatives. Therefore we utilized the more stringent pH of 6.2 for further SAR development. In Tables 1–7 the IC_{50}'s were run at a pH of 4.7 whereas in Tables 9–15 the IC_{50}'s were run at a pH of 6.2.

Stability and protein binding were other issues of concern. After incubating the dihydropyrones in cell culture for 7 days, we were able to recover greater than 90% of the unchanged parent. Obviously, then, the integrity of the sample was not an issue. However, the possibility that the dihydropyrone nucleus would be strongly bound to serum proteins proved more serious. The pharmacokinetic literature indicated that the coumarin template binds very effectively to albumin (Site 1).[54] Our data indicated that the activity of the dihydropyrones in vitro decreased in the presence of albumin. Increasing the polarity of the compounds or raising the pK_a of the 4-hydroxyl should decrease protein binding.

Other conceivable explanations for the lack of cellular activity concerned cellular penetration and active efflux. Our data in the Caco-2 penetration model indicated the compounds should have significant cellular penetration. Although the data from our active efflux studies were not conclusive this also did not appear to be a problem with the compounds as a class.

Table 8. Effect of Assay pH on IC_{50}'s[a]

Compound	R′	R″	IC_{50} at pH 4.7	IC_{50} at pH 6.2
60	$CHMe_2$	H	14	97
64	$CHMe_2$	Me	7	79
63	CMe_3	H	3	104
66	CMe_3	Me	10	35

Note: [a]IC_{50}'s reported in nM.

B. Effect of Polarity on the Cellular Activity

Initial attempts to improve cellular potency centered on the addition of polar functional groups to the dihydropyrone parent. Compound **66** was chosen as the parent structure for these preliminary studies. Examination of the X-ray crystallographic structure of analogue **68** bound to the protease enzyme indicated that several possible sites existed for such synthetic modification. It was hoped that these substitutions could modulate physical properties as well as pick up additional interactions with the protein. With these caveats in mind, several sites for potential modification were identified: the 4' position of the aryl ring at C-3, the 3' or 4' position in the phenethyl ring at C-6, or the 4' position of the phenyl ring at C-6.

The effect of adding a single polar group to the phenethyl moiety at C-6 is summarized in Table 9. In almost all cases, the polar analogues were at least equipotent to the parent compound in the HIV protease assay; the exceptions to this trend were those derivatives containing either a NHAc group (**74**) or two methoxy groups (**76**). Unfortunately, cellular potency for these polar derivatives remained virtually unchanged from the parent—in only one case (**70**, where X = 3-OH) was any hint of cellular activity observed. A similar result was seen when the polar

Table 9. Polar Groups at C-6 Phenethyl

Compound	X	IC_{50} (nM)	EC_{50} (µM)	TC_{50} (µM)
70	3-OH	16	13	57
71	4-CONH$_2$	271	x	x
72	4-NH$_2$	24	>54	54
73	4-OH	11	>69	69
74	4-NHAc	507	x	x
75	3-NH$_2$	17	>65	65
76	3,4-diOCH$_3$	405	x	x
77	3,4-diOH	24	>66	66
78	2-OH	80	>31	>31
66	H	35	>55	55

group was added to the phenyl group at C-6 (Table 10): that is, most analogues retained the desired enzymatic activity but displayed no increased antiviral efficacy. Clearly, though, the flexibility of the dihydropyrone skeleton afforded ample opportunity for substitution and elaboration.

More promising results arose from substitution on the aryl ring at C-3 (Table 11). A variety of substituents were appended to the 4′ position of the S–aryl ring. As noted previously, these polar analogues at least retained the enzyme potency of the parent compound. Furthermore, in four cases—compound **89** (where Z = CH_2OH), compound **88** (where Z = OCH_2CH_2OH), compound **91** (where Z = OCH_2CONH_2), and **93** (where Z = HNAc)—the activity of the polar analogue actually increased three- to fourfold over that of the parent. Unfortunately, this boost in in vitro potency correlated with an increase in cellular potency for **89** only ($EC_{50} = 2.5 \mu M$). For the first time, though, a single polar group conferred an increase in both enzymatic and cellular activity.

The beneficial effects of the CH_2OH moiety warranted further exploration in a series of disubstituted analogues. Introduction of polar functionalities at C-3 and at C-6 resulted in derivatives **95–102** (Table 12). For compounds **95** through **98**, a hydroxyl or amino group was substituted at various positions in the phenethyl ring while the CH_2OH group was kept constant at C-3. All compounds in this series displayed a

Table 10. Polar Groups at C-6 Phenyl

Compound	Y	IC_{50} (nM)	EC_{50} (µM)	TC_{50} (µM)
79	OCH_3	62	>27	27
80	OCH_2CH_2OH	12	9.4	23
81	OCH_2Ph	>196	>30	30
82	OH	40	>23	23
83	NHAc	10	>66	66
84	NH_2	32	>67	67
66	H	35	>55	55

Table 11. Polar Groups at C-3 S-Phenyl

Compound	Z	IC_{50} (nM)	EC_{50} (μM)	TC_{50} (μM)
85	OCH_3	15	>33	33
86	$OCH_2CO_2CH_3$	20	>31	31
87	OH	33	>23	23
88	OCH_2CH_2OH	7	>20	20
89	CH_2OH	7	2.5	66
90	$O(CH_2)_3OH$	18	>25	25
91	OCH_2CONH_2	8	>21	21
92	$OCH_2CONHEt$	38	>64	64
93	NHAc	7	21	69
94	NH_2	11	>64	64
66	H	35	>55	55

Table 12. Polar Groups at C-6 and C-3

Compound	X	Y	Z	IC_{50} (nM)	EC_{50} (μM)	TC_{50} (μM)
95	4-OH	H	CH_2OH	1.7	4.7	>100
96	3-OH	H	CH_2OH	2.5	6.1	>100
97	$4-NH_2$	H	CH_2OH	3.1	3.7	94
98	$3-NH_2$	H	CH_2OH	4.0	3.1	23
99	H	OCH_2CH_2OH	CH_2OH	1.4	4.2	67
100	H	OCH_2CH_2OH	OCH_2CH_2OH	6.4	1.8	25
101	H	OCH_2CH_2OH	OH	3.7	4.2	27
102	4-OH	OCH_3	CH_2OH	x	2.0	>100
66	H	H	H	35	>55	55

dramatic enhancement in enzyme activity; more impressive, however, was the concomitant improvement noted in cellular activity. Every compound in this series possessed low micromolar efficacy in the cellular assay, and in most cases this efficacy was clearly separable from cellular toxicity. For the first time, then, a series of dihydropyrones demonstrated potency in both the enzymatic and cellular assays, producing compounds with therapeutic indices in excess of 20.

For the sake of completeness, a small series of analogues (Table 12, entries **99–101**) was prepared in which a polar group was held constant in the phenyl ring at C-6 while a number of substituents was introduced into the C-3 aryl ring. In this case, the polar group at C-6 was chosen to be OCH_2CH_2OH, a moiety which was previously shown to confer good enzyme activity, albeit without antiviral potency. As before, the enzyme activity of these analogues improved 5- to 25-fold when compared to the parent while the cellular activity reached into the low micromolar range. However, these derivatives also displayed increased toxicity in the cellular assay, especially when compared to analogues in Table 11. It is interesting to note that the addition of the CH_2OH group at C-3 conferred the greatest boost in activity for this disubstituted series, just as it did in the monosubstituted series.

Therefore, addition of the appropriate polar groups to the dihydropyrone framework resulted in marked improvement in both enzymatic and cellular activities. As anticipated, the flexibility of the dihydropyrone skeleton afforded ample opportunity for such modification. In particular, a hydroxyl or amino group on the phenethyl ring at C-6 and a CH_2OH moiety at C-3 seemed to confer antiviral activity without cellular toxicity. At this point, another strategy was devised so that lipophilicity and polarity could be modified without introducing further polar substitution: namely, the phenyl group at C-6 was replaced with various alkyl substituents. Molecular modeling suggested that the pocket at S_1 would easily accommodate the various conformations of different alkyl and cycloalkyl groups.

A series of 6-alkyl derivatives was prepared as summarized in Table 13. In the first series (**103–106**), the phenethyl group was substituted with a 4-OH group and the C-3 aryl moiety contained the benzyl alcohol substituent while the C-6 group was varied. When the aryl group was replaced with the isosteric cyclohexyl moiety (**103**), no significant change in enzyme activity was noted. However, the cellular antiviral potency increased dramatically when compared to the C-6 phenyl analogue (**95**)—a ninefold increase in efficacy. Similar results were obtained

Table 13. 6-Alkyl-6-phenethyl Dihydropyrones

Compound	X	R_6	Z	IC_{50} (nM)	EC_{50} (μM)	TC_{50} (μM)
103	OH	cyclohexyl	CH_2OH	2.5	0.50	75
104	OH	cyclopentyl	CH_2OH	3.1	0.59	>100
105	OH	i-propyl	CH_2OH	3.6	0.53	>100
106	OH	methyl	CH_2OH	4.3	2.5	>100
107	OH	cyclohexyl	NH_2	20	10	66
108	OH	cyclopentyl	NH_2	6.0	2.9	76
109	OH	i-propyl	NH_2	2.7	1.0	92
110	OH	methyl	NH_2	13	3.7	>100
111	NH_2	cyclohexyl	CH_2OH	3.2	1.4	80
112	NH_2	i-propyl	CH_2OH	2.7	0.49	>100

when the alkyl group was changed to cyclopentyl (**104**) and isopropyl (**105**). In addition, these smaller alkyl groups conferred striking enhancements in the toxicity assay (TC_{50}'s > 100 for **104** and **105**). The same general trends hold true for those compounds substituted with a NH_2 group in the C-3 aryl ring (**107–110**): that is, addition of an alkyl group at the C-6 position gives analogues with good antiviral potency, albeit not as potent as the benzyl alcohol comparitors. Potent analogues were also obtained by the substitution of a 4-NH_2 in the C-6 phenethyl ring (**111, 112**). Indeed, derivative **112** possessed low nanomolar potency against the HIV protease enzyme, submicromolar activity in antiviral testing, and low toxicity in the cellular assay. The more potent compounds were screened for PK parameters in mice (vida infra).

C. 6-Alkyl-5,6-Dihydropyran-2-ones Filling the S′₃ Pocket

In an effort to identify more potent inhibitors of HIV PR, a series of compounds was synthesized in which another enzyme pocket could be filled in addition to the four inner pockets occupied by the previously described compounds. This strategy is similar to the one adopted by Upjohn researchers.[32–46] Molecular modeling studies indicated substitution at the 4′-position of the 3-S-(2-*tert*-butyl-5-methyl) phenyl moiety

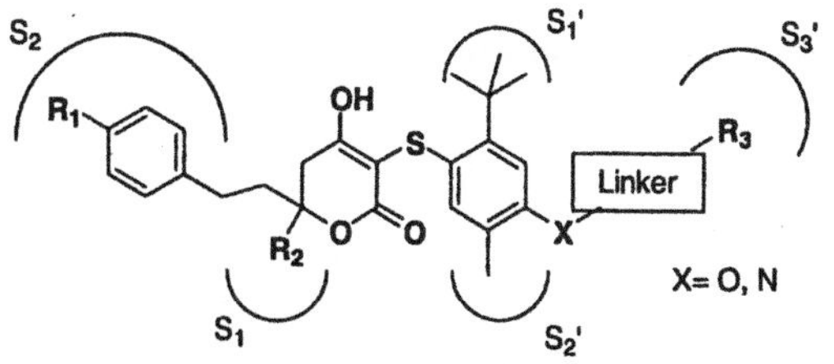

Figure 17. 5,6-Dihydropyrones occupying five enzymatic active-site pockets.

could provide access to the S_3' pocket of the HIV PR (Figure 17). In addition, since the S_3' pocket of the enzyme is located towards the exposed solvent region one could conceivably use various substituents which could both enhance binding affinities and modify physical properties. An example of this strategy involves an amino or hydroxyl group located at the 4'-position of the 3-S-(2-*tert*-butyl-5-methyl) phenyl moiety that could be functionalized to extend into the S_3' pocket of the enzyme. Thus various 5,6-dihydropyran-2-ones possessing carboxamide,[55] sulfamate,[56] urethane, sulfonamide,[57] and sulfonylurea functionalities were synthesized and their activities are shown in Table 14. Among these 5,6-dihydropyran-2-ones (**113–123**), those analogues possessing sulfonamide and sulfamate functionalities exhibited better antiviral activities. The best inhibitor, **120**, possesses an EC_{50} of 220 nM. It is interesting to note that a polar group on the phenethyl moiety in the sulfonamide analogues decreases antiviral activity at least fivefold when compared to the analogue without polar function on phenethyl moiety (**120** vs. **121**), a result that contradicts those observed in the benzyl alcohol series. It is also worth noting that sulfonamide analogues in the Parke-Davis series (containing 3-position sulfur atom) do not enhance either enzymatic binding affinities or antiviral activities significantly, which is contrary to results reported by Upjohn (in a series containing a 3-position carbon atom).

D. Achiral Dihydropyrones

X-ray crystal structures of PD 178390 (**128**, chiral **112**) and PNU-140690 showed similar key interactions with catalytic Asp and the flap Ile. PD 178390 possesses an isopropyl group (occupying the S_1 pocket of the enzyme) and a *p*-amino-phenethyl group (occupying the S_2 pocket of the enzyme) at the 6-position of 5,6-dihydro-pyran-2-one ring (Figure

Table 14. Compounds Designed to Extend into S_3'

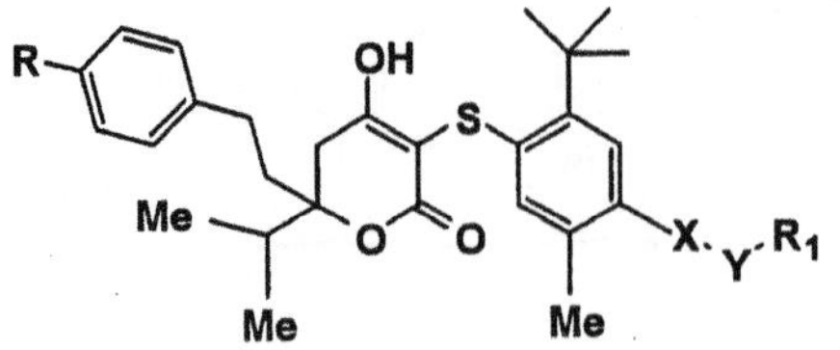

Com-pound	R	X	Y	R_1	Chirality	IC_{50} (nM)	EC_{50} (μM)	TC_{50} (μM)
113	OH	NH$_2$	—	—	S	0.67	0.49	215
114	OH	OH	—	—	RS	0.03	1.6	66
115	OH	NH	CO	Ph(4-CN)	RS	5	4.1	66
116	OH	NH	CO	OtBu	RS	91	—	—
117	OH	NH	SO$_2$	NHEt	RS	1.7	3.4	>100
118	OH	O	SO$_2$	NHEt	RS	4.9	1.0	>100
119	H	NH	SO$_2$	Ph-(4-CN)	RS	9.1	0.52	66
120	H	NH	SO$_2$	Ph-(4-CN)	S	0.20	0.22	31
121	OH	NH	SO$_2$	Ph-(4-CN)	S	2.1	1.6	84
122	OH	NH	SO$_2$	4-CF$_3$-2-Pyr	S	0.30	4.3	66
123	OH	O	SO$_2$	N-MePiperazine	RS	3.1	0.54	>100

18). Structurally, this substitution pattern correlates closely with that of PNU-140690,[45] which contains *n*-propyl and phenethyl groups at the 6-position of the 5,6-dihydropyran-2-one ring. However, the *n*-propyl and phenethyl groups in PNU-140690 occupy the S_2 and S_1 pockets of

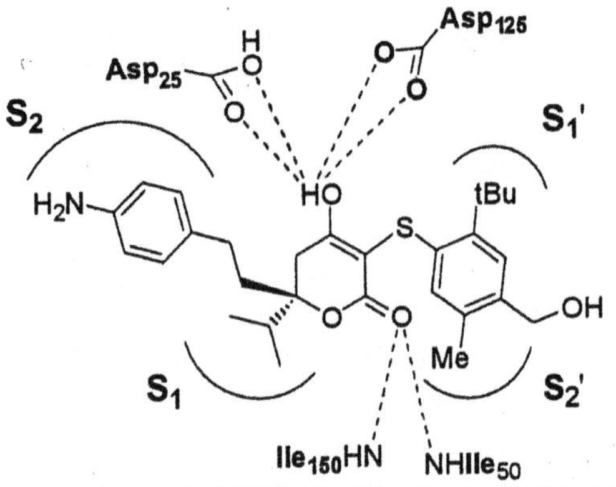

Figure 18. X-ray of PD 178390.

Table 15. Symmetrical Dihydropyrones

Compound	R	IC_{50} (nM)	EC_{50} (M)	TC_{50} (μM)
124	H	150	>23	23
125	p-OH	9	5.8	17
126	p-NH$_2$	10	4.2	24
127	m-OH	16	2.3	17

the enzyme, respectively—the opposite orientation observed in the PD 178390 crystal structure. This result suggests that compounds containing 6,6-dialkyl or 6,6'-diphenethyl groups at the 6-position of 5,6-dihydropyran-2-one ring should be active; moreover, the lack of chirality of these compounds makes them synthetically attractive.

Hamilton et al. has reported a series of 5,6-dihydropyran-2-ones possessing two diphenethyl groups at the 6-position of the ring.[21,22] In this series of compounds (Table 15) the crucial core interactions and the binding at the 3-position are similar to the those described previously for the dihydropyrone class (vide supra). However, one of the phenethyl groups at C-6 was able to extend through the S$_1$ region into S$_3$, and from there into solvent. Despite the attractiveness of these symmetrical targets, the antiviral activities of the compounds are lower than that of the unsymmetrical 6-phenethyl-6-alkyl series.

E. Synthesis of the Dihydropyrones

The synthesis of the racemic dihydropyrones has previously been described.[24,30,31] The chiral synthesis of the dihydropyrones is shown in Scheme 1 as reported by Christopher Gajda.[16,17] The ketone **131** was reacted with the appropriate acetate enolate to afford **132, 134,** or **136**. The resulting ester was converted into the acid **133** by either hydrolyzing the alkyl ester or hydrogenating the benzyl ester. The resulting racemic acid was resolved by classical resolution with a variety of chiral amines to afford resolved **135**. An alternative route to chiral acid **138** was achieved by resolution of the intermediate ester using chiral columns and

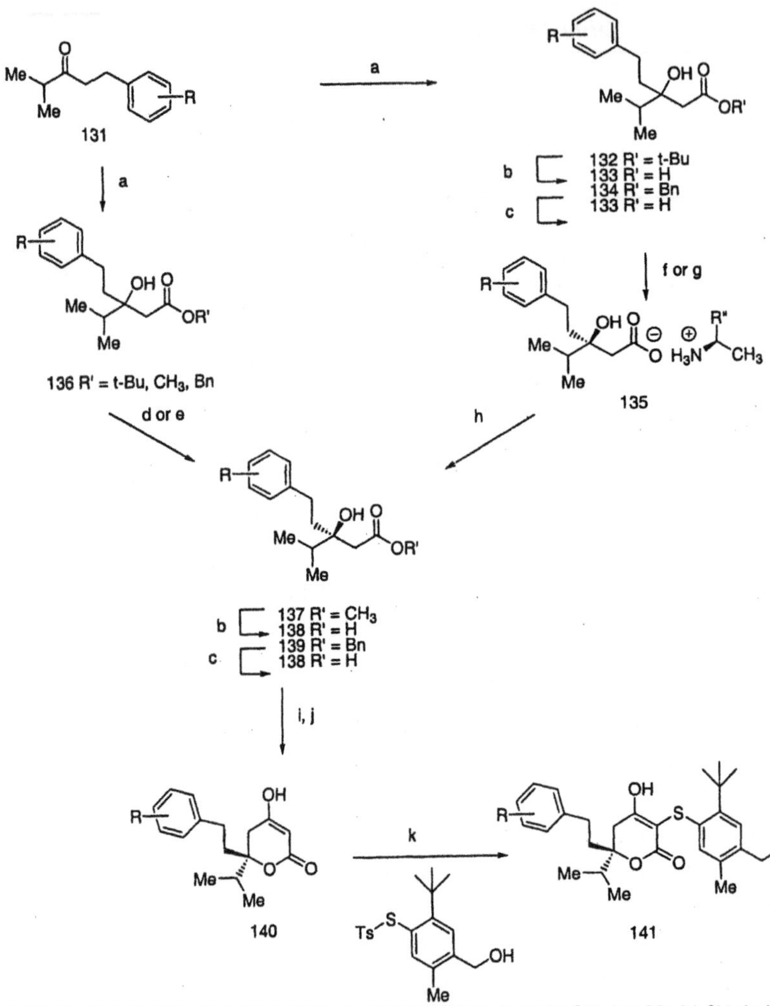

(a) CH_3CO_2R', LDA, THF; (b) LiOH, $MeOH:H_2O$; (c) 20% Pd/C, H_2, THF; (d) Chiralcel OD; (e) Chiralpak AD; (f) (S)-α-methyl naphthylamine, $IPA:H_2O$; (g) (S)-α-methyl benzylamine, EtOAc; (h) 10% HCl; (i) CDI, THF, then $(MeO_2CCH_2CO_2)_2Mg$; (j) 0.1 N NaOH, THF then H^+; (k) K_2CO_3, DMF

Scheme 1. Chiral synthesis of the dihydropyrone template.

then conversion of the chiral ester into the acid. Chiral acid **138** was converted into the final product by conversion to the β-ketoester followed by treatment with base to afford **140**. Reaction of the dihydropyrone with the appropriate tosylate reagent[30,31] in the presence of base afforded the final product **141**.

Table 16. 6-Alkyl-6-phenethyl Dihydropyrones[a]

Compound	X	R_6	Z	EC_{50} (μM)	TC_{50} (μM)	C_{max}
95	OH	Phenyl	CH_2OH	4.7	>100	5.7
103	OH	cyclohexyl	CH_2OH	0.50	75	1.0
104	OH	cyclopentyl	CH_2OH	0.59	>100	6.0
105	OH	i-propyl	CH_2OH	0.53	>100	17
112	NH_2	i-propyl	CH_2OH	0.49	>100	22
128[a]	NH_2	i-propyl	CH_2OH	0.20	>200	40
129	OH	i-propyl	OSO_2-PipMe	0.54	>100	<0.15
130[a]	H	i-propyl	$NHSO_2$-Ph-4-CN	0.22	31	10

Note: [a]All chiral compounds are with the *S* configuration. C_{max} is in mice at 25 mg/kg PO dosed in
solution at pH = 8.0.

F. Pharmacokinetic (PK) Properties

Compounds with a good EC_{50} were further evaluated for plasma levels
and $T_{1/2}$ in mice. Table 16 lists several of the more interesting analogues
tested. As a general rule, we found that as the molecular weight of the
compounds increased, the bioavailability decreased. Interestingly, when
the phenyl group (**95**) was replaced with a cyclohexyl (**103**) the bioavailability decreased dramatically. When the cylclohexyl group (**103**) was
replaced with cyclopentyl (**104**) or isopropyl (**105**) the bioavailability
increased significantly. In general, we found that compounds with bulk-
ier groups extending out into P_3' (**129**, **130**) had less bioavailability than
the CH_2OH compounds (**105**).

G. Profile of PD 178390

A full profile of the best compounds afforded PD 178390 (Table 16,
128) as a compound of interest for further analysis. PD 178390 has one
chiral center and a MW of 483. Figure 18 (vida supra) shows a X-ray of
PD 178390 bound in the active site of the HIV protease enzyme. The
core binds as previously described, the aniline amine binds to the Asp30

via a water bridge and the CH_2OH at the 3-position binds to a backbone NH of the enzyme.

PD 178390 showed good potency in a variety of antiviral assays with an EC_{50} of 0.17 μM, an EC_{90} of 0.41 μM, and TC_{50} of >100 μM in CEM cells against the HIV-1 RF strain. Human serum was added to the antiviral assay to analyze the effect on antiviral activity. The antiviral activity of PD 178390 in H-9 cells with 10% FCS was 0.75 μM and in the presence of 40% human serum the EC_{90} was 3 to 5 μM.

Of vital importance for any new HIV protease inhibitor is its efficacy against resistant strains of HIV. PD 178390 was tested in four low passage PBMC strains and in three clinical strains which were resistant to indinavir (Table 17). We also performed studies on the isolated enzyme as well. In all cases little drop (<10-fold) in the IC_{50}'s was observed with mutated HIV protease.

We further profiled the phamacokinetic properties of PD 178390 in mice, rats, and dogs, (at 25, 10, 10 mg/kg) and found the absolute bioavailability was 63%, 31%, and 43%, respectively. The $T_{1/2}$ was 3.0 h in mice, 4.1 h in rats, and 3.7 h in dogs. Blood levels were maintained above the protein corrected EC_{95} for 12 h in mice, 6 h in rats, and 8 h in dogs.

Efforts to identify potential side effects centered on the question of liver metabolism and enzyme specificity. Most of the approved HIV protease compounds have significant interactions with P450 enzymes, in particular with 3A4. Therefore we evaluated the inhibition of a variety of P450 isozymes with PD 178390. PD 178390 was a very weak inhibitor of all the enzymes tested (EC_{50} > 90 μM against 2D6, 2C9, 3A4, 3A4,

Table 17. Antiviral Activity in PBMC's[a]

	PD 178390
Average of four wild type strains	
EC_{50}	0.54
TD_{50}	>200
Resistant strains	
EC_{50} Strain A	0.20
EC_{50} Strain B	0.20
EC_{50} Strain C	0.56

Note: [a]All data is in μM. Strain A V32I/M46I/L63P/L90M; Strain B L10I/M46I/I54V/L63P/A71V/V82F/L90M; Strain C L10I/M46I/I54V/L63P/V82F/L90M.

1A2, and 2C19). Finally, PD 178390 showed good selectivity for inhibition of HIV PR relative to renin ($EC_{50} > 100$ μM), gastricin ($EC_{50} > 10$ μM), pepsin ($EC_{50} > 10$ μM), Cathepsin D ($EC_{50} > 10$ μM), and Cathepsin E ($EC_{50} > 10$ μM).

VIII. CONCLUSIONS

The utilization of high throughput screening was effective in obtaining an initial hit for further exploration. The tandem of structure-based design and X-ray cocrystallization afforded new ideas and assistance in prioritization of targets, once enzyme potency was optimized. Cellular activity was obtained after the addition of polar functional groups, which decreased protein binding and provided additional interactions with the protein. From this optimization process we identified PD 178390 as a strong preclinical candidate. PD 178390 is a low MW compound with a single chiral center. This agent shows good antiviral activity with little cross-resistance against mutants resistant to the currently approved HIV protease inhibitors. The good bioavailability and half-life of PD 178390 in mice, rats, and dogs indicate few pharmacokinetic problems, while the enzyme selectivity profile bodes well for reduced drug interactions. The combination of excellent antiviral potency, low toxicity, good pharmacokinetic parameters, and promising resistance data make PD 178390 a strong candidate for further development.

ACKNOWLEDGMENTS

The authors would like to first thank the HIV protease team. Unfortunately due to page and time constraints the magnitude of their effort and dedication are not adequately captured in this manuscript. Many thanks to Elizabeth Lunney for her helpful comments and assistance in preparing the color figures and John Domagala for his insightful comments and leadership.

REFERENCES

1. (a) *ASM News* **1998**, *64*, 73. (b) *Hospital Practice* **1996**, *Oct. 15*, 63. (c) Bourinbaiar, Aldar S.; Tan, Xin; Nagomy, Raisa. *AIDS* **1993**, *7*, 129.
2. For general reviews on the functions of HIV protease, see: (a) Darke, P. L.; Huff, J. R. *Adv. Pharmacol.* **1994**, *25*, 399–454. (b) Moyle, G.; Gazzard, B. *Drugs* **1996**, *51*, 701–712. For more clinically-oriented reviews, see: (c) Flexner, C. *Drug Therapy* **1998**, *228(18)*, 1281–1291. (d) Deeks, S. G.; Smith, M.; Holodniy, M.; Kahn, J. O. *JAMA* **1997**, *277*, 145–153.
3. Fitzgerald, Paula M. D. *Curr. Opin. Struct. Biol.* **1993**, *3*, 868–874.

4. Havlir, D. V.; Richman, D. D. *Ann. Intem. Med.* **1996**, *124*, 984–994. For an overview in the popular media, see: Leland, J. *Newsweek*, Dec. 2 1996, 65–73.

5. Barry, M.; Gibbons, S.; Back, D.; Mulcahy, F. *Clin. Pharmacokinetics* **1997**, *32*, 194–209.

6. (a) Mascolini, M. *Journal of the International Assoc. of Physicians in AIDS Care* **1997**, *4*, 23–33. (b) Waldman, A. *Washington Post*, April 27 1997, C01.

7. Van, C.; Gwendolyn, F.; Fisher, E. J.; Polk, R. E. *Pharmacotherapy* **1997**, *17(4)*, 774–778.

8. Wainberg, M.A.; Friedland, G. *JAMA* **1998**, *279(24)*, 1977–1983.

9. Tummino, P. J.; Ferguson, D.; Hupe, D. *Biochem. Biophys. Res. Commun.* **1994**, *201*, 290–294.

10. Tummino, P. J.; Vara Prasad, J. V. N.; Ferguson, D.; Nouhan, C.; Graham, N.; Domagala, J. M.; Ellsworth, E.; Gajda, C.; Hagen, S. E.; Lunney, E. A.; Para, K. S.; Tait, B. D.; Pavlovsky, A.; Erickson, J. W.; Gracheck, S.; McQuade, T. J.; Hupe, D. J. *Bioorg. Med. Chem.* **1996**, *4*, 1401–1410.

11. Para, K. S.; Ellsworth, E. L.; Vara Prasad, J. V. N. *J. Het. Chem.* **1994**, *31*, 1619–1624.

12. Ellsworth, E. L.; Lunney, E.; Sundrum, H.; Tununino, P. ACS Great Lakes Meeting, June 1994, Ann Arbor, MI.

13. Tummino, P. J.; Ferguson, D.; Hupe, L.; Hupe, D. *Biochem. Biophys. Res. Commun.* **1994**, *200*, 1658–1664.

14. Vara Prasad, J. V. N.; Para, K. S.; Lunney, E. A.; Ortwine, D. F.; Dunbar, J. B., Jr.; Ferguson, D.; Tummino, P. J.; Hupe, D.; Tait, B. D.; Domagala, J. M.; Humblet, C.; Bhat, T. N.; Liu, B.; Guerin, D. M. A.; Baldwin, E. T.; Erickson, J. W.; Sawyer, T. K. *J. Amer. Chem.* **1994**, *116*, 6989–6990.

15. Vara Prasad, J. V. N.; Para, K. S.; Tummino, P. J.; Ferguson, D.; McQuade, T. J.; Lunney, E. A.; Rapundalo, S. T.; Batley, B. L.; Hingorani, G.; Domagala, J. M.; Gracheck, S. J.; Bhat, T. N.; Liu, B.; Baldwin, E. T.; Erickson, J. W.; Sawyer, T. K. *J. Med. Chem.* **1995**, *38*, 898–905.

16. Gajda, C.; Domagala, J.; Tait, B.; Hagen, S.; Tummino, P.; Ferguson, D.; Pavlovsky, A.; Rubin, J.; Lunney, E. *210th ACS National Meeting*, Chicago, Illinois, August, 1995.

17. Gajda, C.; Boyer, F. E.; Ellsworth, E. L.; Hagen, S.; Kibbey, C. E.; Lunney, E.; Markoski, L. J.; Pavlovsky, A.; Vara Prasad, J. V. N.; Rubin, J.; Steinbaugh, B.; Tait, B.; Tummino, P.; Urumov, A.; Zeikus, E. *5th Conference on Retroviruses and Opportunistic Infections*, Abst. I-199b, 1998.

18. Stewart, B. H.; Reyner, E. L.; Guttendorf, R. J.; Vara Prasad, J. V. N.; McQuade, T.; Ferguson, D.; Tummino, P.; Tait, B. *Tenth Annual Meeting of Amer. Assoc. of Pharm. Scientists*, San Diego, November, 1995.

19. Tummino, P. J.; Ferguson, D.; Jacobs, C. M.; Tait, B. D.; Hupe, L.; Lunney, E. A.; Hupe, D. *Arch. Biochem. Biophys.* **1995**, *316*, 523–528.

20. Vara Prasad, J. V. N.; Lunney, E. A.; Ferguson, D.; Tummino, P. J.; Rubin, J. R.; Reyner, E. R.; Stewart, B. H.; Guttendorf, R. J.; Domagala, J. M.; Suvorov, L. I.; Gulnik, S. V.; Topol, I. A.; Bhat, T. N.; Erickson, J. W. *J. Am. Chem. Soc.* **1995**, *117*, 11070–11074.

21. Hamilton, H.; Tait, B. D.; Gajda, C.; Hagen, S.; Ferguson, D.; Lunney, E.; Pavlovsky, A.; Tummino, P. *BioOrg. Med. Chem. Lett.* **1996**, *6*, 719–724.

22. Hamilton, H.; Hagen, S.; Steinbaugh, B.; Lunney, E.; Pavlovsky, A.; Domagala, J.; Tummino, P.; Ferguson, D.; Gracheck, S. *XIVth Inter. Symp. on Med. Chem.*, Maastricht, The Netherlands, Sept. 8–12, 1996.

23. Steinbaugh, B. A.; Hamilton, H. W.; Vara Prasad, J. V. N.; Para, K. S.; Tummino, P. J.; Fergusun, D.; Lunney, E. A.; Blankley, C. J. *BioOrg. Med. Chem. Lett.* **1996**, *6*, 1099–1104.

24. Tait, B. D.; Domagala, J.; Ellsworth, E. L.; Ferguson, D.; Gajda, C.; Hupe, D.; Lunney, E. A.; Tummino, P. J. *J. Molec. Recog.* **1996**, *9*, 139–142.

25. Lunney, E. A.; Hagen, S. E.; Domagala, J. M.; Humblet, C.; Kosinski, J.; Tait, B. D.; Warmus, J. S.; Wilson, M.; Ferguson, D.; Hupe, D.; Tummino, P. J.; Baldwin, E. T.; Bhat, T. N.; Liu, B.; Erickson, J. W. *J. Med. Chem.* **1994**, *37*, 2664–2677.

26. Vara Prasad, J. V. N.; Pavlovsky, A.; Para, K. S.; Ellsworth, E. L.; Tummino, P. J.; Nouhan, C.; Ferguson, D. *BioOrg. Med. Chem. Lett.* **1996**, *6*, 1133–1138.

27. Vara Prasad, J. V. N.; Tummino, P. J.; Ferguson, D.; Saunders, J.; VanderRoest, S.; McQuade, T. J.; Sharmeen, L.; Reyner, E. L.; Stewart, B. H.; Para, K. S.; Lunney, E. A.; Gracheck, S. J.; Domagala, J. M. *Biochem. Biophys. Res. Commun.* **1996**, *221*, 815–820.

28. Vara Prasad, J. V. N.; Lunney, E. A.; Para, K. S.; Tummino, P. J.; Ferguson, D.; Hupe, D.; Domagala, J. M.; Erickson, J. W. *Drug Design Disc.* **1996**, *13*, 15–28.

29. Stewart, B. H.; Chung, F. Y.; Tait, B. D.; C. Blankley, C. J.; Chan, O. H. *Pharm. Res.* **1998**, *15*, 1401–1406.

30. Hagen, S.; Vara Prasad, J. V. N.; Boyer, F. E.; Domagala, J. M.; Ellsworth, E. L.; Gajda, C.; Hamilton, H.; Markoski, L.; Steinbaugh, B. A.; Tait, B. D.; Lunney, E. A.; Tummino, P. J.; Ferguson, D.; Hupe, D.; Nouhan, C.; Gracheck, S. J.; Saunders, J. M.; VanderRoest, S. *J. Med. Chem.* **1997**, *40*, 3707–3711.

31. Tait, B. D.; Hagen, S.; Domagala, J.; Ellsworth, E.; Gajda, C.; Hamilton, H.; Vara Prasad, J. V. N.; Ferguson, D.; Hupe, D.; Tummino, P. J.; Humblet, C.; Lunney, E. A.; Pavlovsky, A.; Rubin, J.; Gracheck, S. J.; Baldwin, E. T.; Bhat, T. N.; Erickson, J. W.; Gulnik, S. V.; Beishan L. *J. Med. Chem.* **1997**, *40*, 3781–3792.

32. Thaisrivongs, S.; Tomich, P. K.; Watenpaugh, K. D.; Chong, K-T.; Howe, W. J.; Yang, C-P.; Strohbach, J. W.; Turner, S. R.; McGrath, J. P.; Bohanon, M. J.; Lynn, J. C.; Mulichak, A. M.; Pagano, P. J.; Moon, J. B.; Ruwart, M. J.; Wilkinson, K. F.; Rush, B. D.; Zipp, G. L.; Dalga, R. J.; Schwende, F. J.; Howard, G. M.; Padbury, G. E.; Toth, L. N.; Zhao, Z.; Koeplinger, K. A.; Kakuk, T. J.; Cole, S. L.; Zaya, R. M.; Piper, R. C.; Jeffrey, P. *J. Med. Chem.* **1994**, *37*, 3200–3204.

33. Romines, K. R.; Thaisrivongs, S. *Drugs of the Future* **1995**, *20*, 377.

34. Thaisrivongs, S.; Watenpaugh, K. D.; Howe, W. J.; Tomich, P. K.; Dolak, L. A.; Chong, K-T.; Tomich, C. C.; Tomasselli, A. G.; Turner, S. R.; Strobach, J. W.; Mulichak, A. M.; Janakiraman, M. N.; Moon, J. B.; Lynn, J. C.; Horng, M.; Hinshaw, R. R.; Curry, K. A.; Rothrock, D. J. *J. Med. Chem.* **1995**, *38*, 3624–3637.

35. Romines, K. R.; Watenpaugh, K. D.; Tomich, P. K.; Howe, W. J.; Moris, J. K.; Lovasz, K. D.; Mulichak, A. M.; Finzel, B. C.; Lynn, J. C.; Horng, M.; Schwende, F. J.; Ruwart, M. J.; Zipp, G. L.; Chong, K-T.; Dolak, L. A.; Toth, L. N.; Howard, G. M.; Rush, B. D.; Wilkinson, K. F.; Possert, P. L.; Dalga, R. J.; Hinshaw, R. R. *J. Med. Chem.* **1995**, *38*, 1884–1891.

36. Romines, K. R.; Watenpaugh, K. D.; Howe, W. J.; Tomich, P. K.; Lovasz, K. D.; Morris, J. K.; Janakiraman, M. N.; Lynn, J. C.; Horng, M.; Chong, K-T.; Hinshaw, R. R.; Dolak, L. A. *J. Med. Chem.* **1995**, *38*, 4463–4473.

37. Skulnick, H. I.; Johnson, P. D.; Howe, W. J.; Tomich, P. K.; Chong, K-T.; Watenpaugh, K. D.; Janakiraman, M. N.; Dolak, L. A.; McGrath, J. P.; Lynn, J. C.; Horng, M-M.; Hinshaw, R. R.; Zipp, G. L.; Ruwart, M. J.; Schwende, F. J.; Zhong, W-Z.; Padbury, G. E.; Dalga, R. J.; Shiou, L.; Possert, P. L.; Rush, B. D.; Wilkinson, K. F.; Howard, G. M.; Toth, L. N.; Williams, M. G.; Kakuk, T. J.; Cole, S. L.; Zaya, R. M.; Lovasz, K. D.; Morris, J. K.; Romines, K. R.; Thaisrivongs, S.; Aristoff, P. A. *J. Med. Chem.* **1995**, *38*, 4968–4971.

38. Thaisrivongs, S.; Janakiraman, M. N.; Strobach, Chong, K.-T.; Tomich, P. K.; Dolack, L. A.; Turner, S. R.; Strohbach, J. W.; Lynn, J. C.; Horng, M-M.; Hinshaw R. R.; Watenpaugh, K. D. *J. Med. Chem.* **1996**, *39*, 2400–2410.

39. Romines, K. R.; Morris, J. K.; Howe, W. J.; Tomich, P. K.; Horng, M-M.; Chong, K-T.; Hinshaw, R. R.; Anderson, D. J.; Strohbach, J. W.; Turner, S. R.; Mizsak, S. A. *J. Med. Chem.* **1996**, *39*, 4125–4130.

40. Thaisrivongs, S.; Skulnick, H. I.; Turner, S. R.; Strohbach, J. W.; Tommasi, R. A.; Johnson, P. D.; Aristoff, P. A.; Judge, T. M.; Gammill, R. B.; Morris, J. K.; Romines, K. R.; Chrusciel, R. A.; Hinshaw, R. R.; Chong, K-T.; Tarpley, W. G.; Poppe, S. M.; Slade, D. E.; Lynn, J. C.; Horng, M-M.; Tomich, P. K.; Seest, E. P.; Dolak, L. A.; Howe, W. J.; Howard, G. M.; Schwende, F. J.; Toth, L. N.; Padbury, G. E.; Wilson, G. J.; Shiou, L.; Zipp, G. L.; Wilkinson, K. F.; Rush, B. D.; Ruwart, M. J.; Koeplinger, K. A.; Zhao, Z.; Cole, S.; Zaya, R. M.; Kakuk, T. J.; Janakiraman, M. N.; Watenpaugh, K. D. *J. Med. Chem.* **1996**, *39*, 4349–4353.

41. Thaisrivongs, S.; Romero, D. L.; Tommasi, R. A.; Janakiraman, M. N.; Strohbach, J. W.; Turner, S. R.; Biles, C.; Morge, R. R.; Johnson, P. D.; Aristoff, P. A.; Tomich, P. K.; Lynn, J. C.; Horng, M-M.; Chong, K.-T.; Hinshaw, R. R.; Howe, W. J.; Finzel, B. C.; Watenpaugh, K. D. *J. Med. Chem.* **1996**, *39*, 4630–4642.

42. Skulnick, H. I.; et. al. *J. Med. Chem.* **1997**, *40*, 1149–1164.

43. Thaisrivongs, S.; Skulnick, H. I.; Turner, S. R.; Strobach, J. W.; Tommasi, R. A.; Johnson, P. D.; Aristoff, P. A.; Judge, T. M.; Ganunill, R. B.; Morris, J. K.; Romines, K. R.; Chrusciel, R. A.; Hinshaw, R. R.; Chong, K.-T.; Tarpley, W. G.; Poppe, S. M.; Slade, D. E.; Lynn, J. C.; Horng, M.-M.; Tomich, P. K.; Seest, E. P.; Dolak, L. A.; Howe, W. J.; Howard, G. M.; Schwende, F. J.; Toth, L. N.; Padbury, G. E.; Wilson, K. F.; Rush, B. D.; Ruwart, M. J.; Koeplinger, K. A.; Zhao, Z.; Cole, S.; Zaya, R. M.; Kakuk, T. J.; Janakiraman, M. N.; Watenpaugh, K. D. *J. Med. Chem.* **1996**, *39*, 4349–4353.

44. Janakiraman, M. N.; Watenpaugh, K. D.; Tomich, P. K.; Chong, K-T.; Turner, S. R.; Tommasi, R. A.; Thaisrivongs, S.; Strohbach, J. W. *Bioorg. Med. Chem. Let.* **1998**, *8*, 1237–1242.

45. PNU-140690 *Drugs of the Future* **1998**, *23*, 146.

46. Turner, S. R.; Strobach, J. W.; Tommasi, R. A.; Aristoff, P. A.; Johnson, P. D.; Skulnick, H. I.; Dolak, L. A.; Seest, E. P.; Tomich, P. K.; Bohanon, M. J.; Horng, M-A.; Lynn, J. C.; Chong, K-T.; Hinshaw, R. R.; Watenpaugh, K. D.; Janikiraman, M. N.; Tharsiivongs, S. *J. Med. Chem.* **1998**, *41*, 3467–3476.

47. Bourinbaiar, A. S.; Tan, X.; Nagorny, R. *Acta. Virol.* **1193**, *37*, 241.

48. Luly, J. R.; Yi, N.; Soderquist, J.; Stein, H.; Cohen, J.; Perun, T. J.; Plattner, J. J. *J. Med. Chem.* **1987**, *30*, 1609–1616.
49. Jungheim, L. N.; Shepherd, T. A.; Baxter, A. J.; Burgess, J.; Hatch, S. D.; Lubbehusen, P.; Wiskerchen, M.; Muesing, M. A. *J. Med. Chem.* **1996**, *39*, 96–108 and references therein.
50. Topliss, J. G.; Martin, Y. C. *Drug Design*; Ariens, E. J., Ed.; Academic Press: New York, 1975, Vol. 5, pp. 1–21.
51. Thomson, W. J.; Fitzgerald, P. M. D.; Holloway, K.; Emini, E. A.; Darke, P. L.; McKeever, B. M.; Schleif, W. A.; Quintero, J. C.; Zugay, J. A.; Tucker, T. J.; Schwering, J. E.; Hommick, C. F.; Nunberg, J.; Springer, J. P.; Huff, J. R. *J. Med. Chem.* **1992**, *35*, 1685–1701.
52. Kempf, D. J.; Codacovi, L.; Wang, X. C.; Kohlbrenner, W. E.; Wideburg, N. E.; Saldivar, A.; Vasavanonda, S.; Marsh, K. C.; Bryant, P.; Sham, H. L.; Green, B. E.; Betebenner, D. A.; Erickson, J.; Norbeck, D. W. *J. Med. Chem.* **1993**, *36*, 320–330.
53. Rich, D. H.; Vara Prasad, J. V. N.; Sun, C. Q.; Green, J.; Mueller, R. D.; Mckenzie, D.; Malkovsky, M. *J. Med. Chem.* **1992**, *35*, 3803–3812.
54. Sudlow, G.; Birkett, D. J.; Wade, D. N. *Molec. Pharmacol.* **1976**, *12*, 1052–1061.
55. Vara Prasad, J. V. N.; Boyer, F. E.; Domagala, J. M.; Ellsworth, E. L.; Gajda, C.; Hagen, S. E.; Markoski, L. J.; Tait, B. D.; Lunney, E. A.; Tummino, P. J.; Ferguson, D.; Holler, T.; Hupe, D.; Nouhan, C.; Gracheek, S. J.; VanderRoest, S.; Saunders, J. M.; Iyer, K.; Sinz, M.; Brodfuehrer, J. *BioOrg. Med. Chem. Lett.* **1999**. In press.
56. Vara Prasad, J. V. N.; Markoski, L. J.; Boyer, F. E.; Domagala, J. M.; Ellsworth, E. L.; Gajda, C.; Hagen, S. E.; Tait, B. D.; Lunney, E. A.; Tummino, P. J.; Ferguson, D.; Holler, T.; Hupe, D.; Nouhan, C.; Gracheek, S. J.; VanderRoest, S.; Saunders, J. M.; Iyer, K.; Sinz, M. Private communication.
57. Boyer, F. E.; Vara Prasad, J. V. N.; Domagala, J. M.; Ellsworth, E. L.; Gajda, C.; Hagen, S. E.; Markoski, L. J.; Tait, B. D.; Lunney, E. A.; Pavlovsky, A.; Ferguson, D.; Graham, N.; Holler, T.; Hupe, D.; Nouhan, C.; Tummino, P. J.; Urumov, A.; Zeikus, E.; Zeikus, G.; Gracheek, S. J.; Saunders, L. M.; VanderRoest, S.; Brodfuehrer, J.; Iyer, K.; Sinz, M.; Gulnik, S. V.; Erickson, J. W. Private communication.

INDEX

Printed and bound by CPI Group (UK) Ltd, Croydon, CR0 4YY

13/10/2024

01773500-0002